FRONTIERS IN APPLIED MATHEMATICS

The SIAM series on Frontiers in Applied Mathematics publishes monographs dealing with creative work in a substantive field involving applied mathematics or scientific computation. All works focus on emerging or rapidly developing research areas that report on new techniques to solve mainstream problems in science or engineering.

The goal of the series is to promote, through short, inexpensive, expertly written monographs, cutting edge research poised to have a substantial impact on the solutions of problems that advance science and technology. The volumes encompass a broad spectrum of topics important to the applied mathematical areas of education, government, and industry.

EDITORIAL BOARD

BOOKS PUBLISHED IN FRONTIERS IN APPLIED MATHEMATICS

Kelley, C. T., *Iterative Methods for Optimization*

Greenbaum, Anne, *Iterative Methods for Solving Linear Systems*

Kelley, C. T., *Iterative Methods for Linear and Nonlinear Equations*

Bank, Randolph E., *PLTMG: A Software Package for Solving Elliptic Partial Differential Equations. Users' Guide 7.0*

Moré, Jorge J. and Wright, Stephen J., *Optimization Software Guide*

Rüde, Ulrich, *Mathematical and Computational Techniques for Multilevel Adaptive Methods*

Cook, L. Pamela, *Transonic Aerodynamics: Problems in Asymptotic Theory*

Banks, H. T., *Control and Estimation in Distributed Parameter Systems*

Van Loan, Charles, *Computational Frameworks for the Fast Fourier Transform*

Van Huffel, Sabine and Vandewalle, Joos, *The Total Least Squares Problem: Computational Aspects and Analysis*

Castillo, José E., *Mathematical Aspects of Numerical Grid Generation*

Bank, R. E., *PLTMG: A Software Package for Solving Elliptic Partial Differential Equations. Users' Guide 6.0*

McCormick, Stephen F., *Multilevel Adaptive Methods for Partial Differential Equations*

Grossman, Robert, *Symbolic Computation: Applications to Scientific Computing*

Coleman, Thomas F. and Van Loan, Charles, *Handbook for Matrix Computations*

McCormick, Stephen F., *Multigrid Methods*

Buckmaster, John D., *The Mathematics of Combustion*

Ewing, Richard E., *The Mathematics of Reservoir Simulation*

Iterative Methods for Optimization

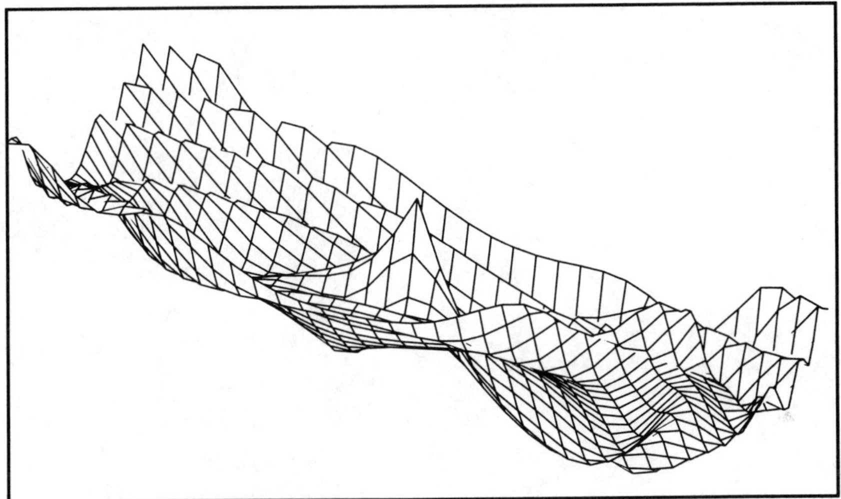

Iterative Methods for Optimization

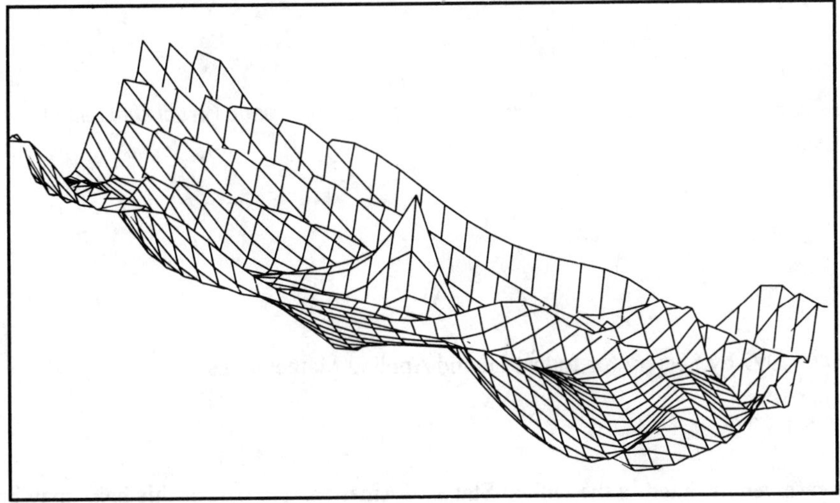

C. T. Kelley

North Carolina State University
Raleigh, North Carolina

siam.

Society for Industrial and Applied Mathematics
Philadelphia

10 9 8 7 6 5 4 3 2

Library of Congress Cataloging-in-Publication Data

Kelley, C. T.
Iterative methods for optimization / C.T. Kelley.
p. cm. -- (Frontiers in applied mathematics)
Includes bibliographical references and index.
ISBN 0-89871-433-8 (pbk.)
1. Mathematical optimization. 2. Iterative methods (Mathematics) I. Title. II. Series.
QA402.5.K44 1999
519.3--dc21 99-11141

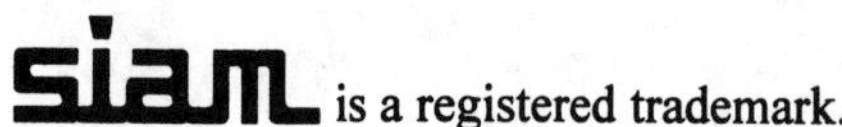

To Chung-Wei and my parents

Contents

Preface

This book on unconstrained and bound constrained optimization can be used as a tutorial for self-study or a reference by those who solve such problems in their work. It can also serve as a textbook in an introductory optimization course.

As in my earlier book [154] on linear and nonlinear equations, we treat a small number of methods in depth, giving a less detailed description of only a few (for example, the nonlinear conjugate gradient method and the DIRECT algorithm). We aim for clarity and brevity rather than complete generality and confine our scope to algorithms that are easy to implement (by the reader!) and understand.

One consequence of this approach is that the algorithms in this book are often special cases of more general ones in the literature. For example, in Chapter 3, we provide details only for trust region globalizations of Newton's method for unconstrained problems and line search globalizations of the BFGS quasi-Newton method for unconstrained and bound constrained problems. We refer the reader to the literature for more general results. Our intention is that both our algorithms and proofs, being special cases, are more concise and simple than others in the literature and illustrate the central issues more clearly than a fully general formulation.

Part II of this book covers some algorithms for noisy or global optimization or both. There are many interesting algorithms in this class, and this book is limited to those deterministic algorithms that can be implemented in a more-or-less straightforward way. We do not, for example, cover simulated annealing, genetic algorithms, response surface methods, or random search procedures.

The reader of this book should be familiar with the material in an elementary graduate level course in numerical analysis, in particular direct and iterative methods for the solution of linear equations and linear least squares problems. The material in texts such as [127] and [264] is sufficient.

A suite of MATLAB* codes has been written to accompany this book. These codes were used to generate the computational examples in the book, but the algorithms do not depend on the MATLAB environment and the reader can easily implement the algorithms in another language, either directly from the algorithmic descriptions or by translating the MATLAB code. The MATLAB environment is an excellent choice for experimentation, doing the exercises, and small-to-medium-scale production work. Large-scale work on high-performance computers is best done in another language. The reader should also be aware that there is a large amount of high-quality software available for optimization. The book [195], for example, provides pointers to several useful packages.

Parts of this book are based upon work supported by the National Science Foundation over several years, most recently under National Science Foundation grants DMS-9321938, DMS-9700569, and DMS-9714811, and by allocations of computing resources from the North Carolina Supercomputing Center. Any opinions, findings, and conclusions or recommendations expressed

*MATLAB is a registered trademark of The MathWorks, Inc., 24 Prime Park Way, Natick, MA 01760, USA, (508) 653-1415, info@mathworks.com, http://www.mathworks.com.

in this material are those of the author and do not necessarily reflect the views of the National Science Foundation or of the North Carolina Supercomputing Center.

The list of students and colleagues who have helped me with this project, directly, through collaborations/discussions on issues that I treat in the manuscript, by providing pointers to the literature, or as a source of inspiration, is long. I am particularly indebted to Tom Banks, Jim Banoczi, John Betts, David Bortz, Steve Campbell, Tony Choi, Andy Conn, Douglas Cooper, Joe David, John Dennis, Owen Eslinger, Jörg Gablonsky, Paul Gilmore, Matthias Heinkenschloß, Laura Helfrich, Lea Jenkins, Vickie Kearn, Carl and Betty Kelley, Debbie Lockhart, Casey Miller, Jorge Moré, Mary Rose Muccie, John Nelder, Chung-Wei Ng, Deborah Poulson, Ekkehard Sachs, Dave Shanno, Joseph Skudlarek, Dan Sorensen, John Strikwerda, Mike Tocci, Jon Tolle, Virginia Torczon, Floria Tosca, Hien Tran, Margaret Wright, Steve Wright, and Kevin Yoemans.

C. T. Kelley
Raleigh, North Carolina

How to Get the Software

All computations reported in this book were done in MATLAB (version 5.2 on various SUN SPARCstations and on an Apple Macintosh Powerbook 2400). The suite of MATLAB codes that we used for the examples is available by anonymous ftp from ftp.math.ncsu.edu in the directory

`FTP/kelley/optimization/matlab`

or from SIAM's World Wide Web server at

`http://www.siam.org/books/fr18/`

One can obtain MATLAB from
The MathWorks, Inc.
3 Apple Hill Drive
Natick, MA 01760-2098
(508) 647-7000
Fax: (508) 647-7001
E-mail: info@mathworks.com
WWW: http://www.mathworks.com

Part I

Optimization of Smooth Functions

Chapter 1

Basic Concepts

1.1 The Problem

The unconstrained optimization problem is to minimize a real-valued function f of N variables. By this we mean to find a *local minimizer*, that is, a point x^* such that

$$f(x^*) \leq f(x) \text{ for all } x \text{ near } x^*. \tag{1.1}$$

It is standard to express this problem as

$$\min_x f(x) \tag{1.2}$$

or to say that we seek to solve the problem $\min f$. The understanding is that (1.1) means that we seek a local minimizer. We will refer to f as the *objective function* and to $f(x^*)$ as the *minimum* or *minimum value*. If a local minimizer x^* exists, we say a *minimum is attained* at x^*.

We say that problem (1.2) is *unconstrained* because we impose no conditions on the independent variables x and assume that f is defined for all x.

The local minimization problem is different from (and much easier than) the *global minimization problem* in which a *global minimizer*, a point x^* such that

$$f(x^*) \leq f(x) \text{ for all } x, \tag{1.3}$$

is sought.

The *constrained* optimization problem is to minimize a function f over a set $U \subset R^N$. A local minimizer, therefore, is an $x^* \in U$ such that

$$f(x^*) \leq f(x) \text{ for all } x \in U \text{ near } x^*. \tag{1.4}$$

Similar to (1.2) we express this as

$$\min_{x \in U} f(x) \tag{1.5}$$

or say that we seek to solve the problem $\min_U f$. A global minimizer is a point $x^* \in U$ such that

$$f(x^*) \leq f(x) \text{ for all } x \in U. \tag{1.6}$$

We consider only the simplest constrained problems in this book (Chapter 5 and §7.4) and refer the reader to [104], [117], [195], and [66] for deeper discussions of constrained optimization and pointers to software.

Having posed an optimization problem one can proceed in the classical way and use methods that require smoothness of f. That is the approach we take in this first part of the book. These methods can fail if the objective function has discontinuities or irregularities. Such nonsmooth effects are common and can be caused, for example, by truncation error in internal calculations for f, noise in internal probabilistic modeling in f, table lookup within f, or use of experimental data in f. We address a class of methods for dealing with such problems in Part II.

1.2 Notation

In this book, following the convention in [154], vectors are to be understood as column vectors. The vector x^* will denote a solution, x a potential solution, and $\{x_k\}_{k\geq 0}$ the sequence of iterates. We will refer to x_0 as the *initial iterate*. x_0 is sometimes timidly called the *initial guess*. We will denote the ith component of a vector x by $(x)_i$ (note the parentheses) and the ith component of x_k by $(x_k)_i$. We will rarely need to refer to individual components of vectors. We will let $\partial f/\partial x_i$ denote the partial derivative of f with respect to $(x)_i$. As is standard [154], $e = x - x^*$ will denote the error, $e_n = x_n - x^*$ the error in the nth iterate, and $\mathcal{B}(r)$ the ball of radius r about x^*

$$\mathcal{B}(r) = \{x \mid \|e\| < r\}.$$

For $x \in R^N$ we let $\nabla f(x) \in R^N$ denote the *gradient* of f,

$$\nabla f(x) = (\partial f/\partial x_1, \ldots, \partial f/\partial x_N),$$

when it exists.

We let $\nabla^2 f$ denote the *Hessian* of f,

$$(\nabla^2 f)_{ij} = \partial^2 f/\partial x_i \partial x_j,$$

when it exists. Note that $\nabla^2 f$ is the Jacobian of ∇f. However, $\nabla^2 f$ has more structure than a Jacobian for a general nonlinear function. If f is twice continuously differentiable, then the Hessian is symmetric ($(\nabla^2 f)_{ij} = (\nabla^2 f)_{ji}$) by equality of mixed partial derivatives [229].

In this book we will consistently use the Euclidean norm

$$\|x\| = \sqrt{\sum_{i=1}^{N} (x)_i^2}.$$

When we refer to a matrix norm we will mean the matrix norm induced by the Euclidean norm

$$\|A\| = \max_{x \neq 0} \frac{\|Ax\|}{\|x\|}.$$

In optimization definiteness or semidefiniteness of the Hessian plays an important role in the necessary and sufficient conditions for optimality that we discuss in §1.3 and 1.4 and in our choice of algorithms throughout this book.

DEFINITION 1.2.1. *An $N \times N$ matrix A is* positive semidefinite *if $x^T Ax \geq 0$ for all $x \in R^N$. A is* positive definite *if $x^T Ax > 0$ for all $x \in R^N, x \neq 0$. If A has both positive and negative eigenvalues, we say A is* indefinite. *If A is symmetric and positive definite, we will say A is* spd.

We will use two forms of the *fundamental theorem of calculus*, one for the function–gradient pair and one for the gradient–Hessian.

THEOREM 1.2.1. *Let f be twice continuously differentiable in a neighborhood of a line segment between points $x^*, x = x^* + e \in R^N$; then*

$$f(x) = f(x^*) + \int_0^1 \nabla f(x^* + te)^T e \, dt$$

and

$$\nabla f(x) = \nabla f(x^*) + \int_0^1 \nabla^2 f(x^* + te)e\,dt.$$

A direct consequence (see Exercise 1.7.1) of Theorem 1.2.1 is the following form of *Taylor's theorem* we will use throughout this book.

THEOREM 1.2.2. *Let f be twice continuously differentiable in a neighborhood of a point $x^* \in R^N$. Then for $e \in R^N$ and $\|e\|$ sufficiently small*

$$f(x^* + e) = f(x^*) + \nabla f(x^*)^T e + e^T \nabla^2 f(x^*) e/2 + o(\|e\|^2). \tag{1.7}$$

1.3 Necessary Conditions

Let f be twice continuously differentiable. We will use Taylor's theorem in a simple way to show that the gradient of f vanishes at a local minimizer and the Hessian is positive semidefinite. These are the *necessary conditions* for optimality.

The necessary conditions relate (1.1) to a nonlinear equation and allow one to use fast algorithms for nonlinear equations [84], [154], [211] to compute minimizers. Therefore, the necessary conditions for optimality will be used in a critical way in the discussion of local convergence in Chapter 2. A critical first step in the design of an algorithm for a new optimization problem is the formulation of necessary conditions. Of course, the gradient vanishes at a maximum, too, and the utility of the nonlinear equations formulation is restricted to a neighborhood of a minimizer.

THEOREM 1.3.1. *Let f be twice continuously differentiable and let x^* be a local minimizer of f. Then*

$$\nabla f(x^*) = 0.$$

Moreover $\nabla^2 f(x^*)$ is positive semidefinite.

Proof. Let $u \in R^N$ be given. Taylor's theorem states that for all real t sufficiently small

$$f(x^* + tu) = f(x^*) + t\nabla f(x^*)^T u + \frac{t^2}{2} u^T \nabla^2 f(x^*) u + o(t^2).$$

Since x^* is a local minimizer we must have for t sufficiently small $0 \le f(x^* + tu) - f(x^*)$ and hence

$$\nabla f(x^*)^T u + \frac{t}{2} u^T \nabla^2 f(x^*) u + o(t) \ge 0 \tag{1.8}$$

for all t sufficiently small and all $u \in R^N$. So if we set $t = 0$ and $u = -\nabla f(x^*)$ we obtain

$$\|\nabla f(x^*)\|^2 = 0.$$

Setting $\nabla f(x^*) = 0$, dividing by t, and setting $t = 0$ in (1.8), we obtain

$$\frac{1}{2} u^T \nabla^2 f(x^*) u \ge 0$$

for all $u \in R^N$. This completes the proof. □

The condition that $\nabla f(x^*) = 0$ is called the *first-order necessary condition* and a point satisfying that condition is called a *stationary point* or a *critical point*.

1.4 Sufficient Conditions

A stationary point need not be a minimizer. For example, the function $\phi(t) = -t^4$ satisfies the necessary conditions at 0, which is a maximizer of ϕ. To obtain a minimizer we must require that the second derivative be nonnegative. This alone is not sufficient (think of $\phi(t) = t^3$) and only if the second derivative is strictly positive can we be completely certain. These are the *sufficient conditions* for optimality.

THEOREM 1.4.1. *Let f be twice continuously differentiable in a neighborhood of x^*. Assume that $\nabla f(x^*) = 0$ and that $\nabla^2 f(x^*)$ is positive definite. Then x^* is a local minimizer of f.*

Proof. Let $0 \neq u \in R^N$. For sufficiently small t we have

$$\begin{aligned} f(x^* + tu) &= f(x^*) + t\nabla f(x^*)^T u + \frac{t^2}{2} u^T \nabla^2 f(x^*) u + o(t^2) \\ &= f(x^*) + \frac{t^2}{2} u^T \nabla^2 f(x^*) u + o(t^2). \end{aligned}$$

Hence, if $\lambda > 0$ is the smallest eigenvalue of $\nabla^2 f(x^*)$ we have

$$f(x^* + tu) - f(x^*) \geq \frac{\lambda}{2} \|tu\|^2 + o(t^2) > 0$$

for t sufficiently small. Hence x^* is a local minimizer. □

1.5 Quadratic Objective Functions

The simplest optimization problems are those with *quadratic objective functions*. Here

$$f(x) = -x^T b + \frac{1}{2} x^T H x. \tag{1.9}$$

We may, without loss of generality, assume that H is symmetric because

$$x^T H x = x^T \left(\frac{H + H^T}{2} \right) x. \tag{1.10}$$

Quadratic functions form the basis for most of the algorithms in Part I, which approximate an objective function f by a *quadratic model* and minimize that model. In this section we discuss some elementary issues in quadratic optimization.

Clearly,

$$\nabla^2 f(x) = H$$

for all x. The symmetry of H implies that

$$\nabla f(x) = -b + Hx.$$

DEFINITION 1.5.1. *The quadratic function f in* (1.9) *is* convex *if H is positive semidefinite.*

1.5.1 Positive Definite Hessian

The necessary conditions for optimality imply that if a quadratic function f has a local minimum x^*, then H is positive semidefinite and

$$Hx^* = b. \tag{1.11}$$

In particular, if H is spd (and hence nonsingular), the unique global minimizer is the solution of the linear system (1.11).

If H is a dense matrix and N is not too large, it is reasonable to solve (1.11) by computing the *Cholesky factorization* [249], [127] of H

$$H = LL^T,$$

where L is a nonsingular lower triangular matrix with positive diagonal, and then solving (1.11) by two triangular solves. If H is indefinite the Cholesky factorization will not exist and the standard implementation [127], [249], [264] will fail because the computation of the diagonal of L will require a real square root of a negative number or a division by zero.

If N is very large, H is sparse, or a matrix representation of H is not available, a more efficient approach is the *conjugate gradient* iteration [154], [141]. This iteration requires only matrix–vector products, a feature which we will use in a direct way in §§2.5 and 3.3.7. Our formulation of the algorithm uses x as both an input and output variable. On input x contains x_0, the initial iterate, and on output the approximate solution. We terminate the iteration if the relative residual is sufficiently small, i.e.,

$$\|b - Hx\| \leq \epsilon \|b\|$$

or if too many iterations have been taken.

ALGORITHM 1.5.1. $\texttt{cg}(x, b, H, \epsilon, kmax)$

1. $r = b - Hx$, $\rho_0 = \|r\|^2$, $k = 1$.
2. *Do While* $\sqrt{\rho_{k-1}} > \epsilon\|b\|$ *and* $k < kmax$
   (a) *if* $k = 1$ *then* $p = r$
   else
   $\beta = \rho_{k-1}/\rho_{k-2}$ *and* $p = r + \beta p$
   (b) $w = Hp$
   (c) $\alpha = \rho_{k-1}/p^T w$
   (d) $x = x + \alpha p$
   (e) $r = r - \alpha w$
   (f) $\rho_k = \|r\|^2$
   (g) $k = k + 1$

Note that if H is not spd, the denominator in $\alpha = \rho_{k-1}/p^T w$ may vanish, resulting in *breakdown* of the iteration.

The conjugate gradient iteration minimizes f over an increasing sequence of nested subspaces of R^N [127], [154]. We have that

$$f(x_k) \leq f(x) \text{ for all } x \in x_0 + \mathcal{K}_k,$$

where $\mathcal{K}_k$ is the *Krylov subspace*

$$\mathcal{K}_k = \text{span}(r_0, Hr_0, \ldots, H^{k-1}r_0)$$

for $k \geq 1$.

While in principle the iteration must converge after N iterations and conjugate gradient can be regarded as a direct solver, N is, in practice, far too many iterations for the large problems to which conjugate gradient is applied. As an iterative method, the performance of the conjugate gradient algorithm depends both on b and on the spectrum of H (see [154] and the references cited therein). A general convergence estimate [68], [60], which will suffice for the discussion here, is

$$\|x_k - x^*\|_H \leq 2\|x_0 - x^*\|_H \left[\frac{\sqrt{\kappa(H)} - 1}{\sqrt{\kappa(H)} + 1}\right]^k. \tag{1.12}$$

In (1.12), the H-norm of a vector is defined by

$$\|u\|_H^2 = u^T Hu$$

for an spd matrix H. $\kappa(H)$ is the l^2 condition number

$$\kappa(H) = \|H\|\|H^{-1}\|.$$

For spd H

$$\kappa(H) = \lambda_l \lambda_s^{-1},$$

where λ_l and λ_s are the largest and smallest eigenvalues of H. Geometrically, $\kappa(H)$ is large if the ellipsoidal level surfaces of f are very far from spherical.

The conjugate gradient iteration will perform well if $\kappa(H)$ is near 1 and may perform very poorly if $\kappa(H)$ is large. The performance can be improved by *preconditioning*, which transforms (1.11) into one with a coefficient matrix having eigenvalues near 1. Suppose that M is spd and a sufficiently good approximation to H^{-1} so that

$$\kappa(M^{1/2}HM^{1/2})$$

is much smaller that $\kappa(H)$. In that case, (1.12) would indicate that far fewer conjugate gradient iterations might be needed to solve

$$M^{1/2}HM^{1/2}z = M^{1/2}b \tag{1.13}$$

than to solve (1.11). Moreover, the solution x^* of (1.11) could be recovered from the solution z^* of (1.13) by

$$x = M^{1/2}z. \tag{1.14}$$

In practice, the square root of the preconditioning matrix M need not be computed. The algorithm, using the same conventions that we used for cg, is

ALGORITHM 1.5.2. pcg$(x, b, H, M, \epsilon, kmax)$

1. $r = b - Hx$, $\rho_0 = \|r\|^2$, $k = 1$
2. *Do While* $\sqrt{\rho_{k-1}} > \epsilon\|b\|$ *and* $k < kmax$
    (a) $z = Mr$
    (b) $\tau_{k-1} = z^T r$

(c) *if* $k = 1$ *then* $\beta = 0$ *and* $p = z$
else
$\beta = \tau_{k-1}/\tau_{k-2}, p = z + \beta p$

(d) $w = Hp$

(e) $\alpha = \tau_{k-1}/p^T w$

(f) $x = x + \alpha p$

(g) $r = r - \alpha w$

(h) $\rho_k = r^T r$

(i) $k = k + 1$

Note that only products of M with vectors in R^N are needed and that a matrix representation of M need not be stored. We refer the reader to [11], [15], [127], and [154] for more discussion of preconditioners and their construction.

1.5.2 Indefinite Hessian

If H is indefinite, the necessary conditions, Theorem 1.3.1, imply that there will be no local minimum. Even so, it will be important to understand some properties of quadratic problems with indefinite Hessians when we design algorithms with initial iterates far from local minimizers and we discuss some of the issues here.

If

$$u^T H u < 0,$$

we say that u is a *direction of negative curvature*. If u is a direction of negative curvature, then $f(x + tu)$ will decrease to $-\infty$ as $t \to \infty$.

1.6 Examples

It will be useful to have some example problems to solve as we develop the algorithms. The examples here are included to encourage the reader to experiment with the algorithms and play with the MATLAB codes. The codes for the problems themselves are included with the set of MATLAB codes. The author of this book does not encourage the reader to regard the examples as anything more than examples. In particular, they are not real-world problems, and should not be used as an exhaustive test suite for a code. While there are documented collections of test problems (for example, [10] and [26]), the reader should always evaluate and compare algorithms in the context of his/her own problems.

Some of the problems are directly related to applications. When that is the case we will cite some of the relevant literature. Other examples are included because they are small, simple, and illustrate important effects that can be hidden by the complexity of more serious problems.

1.6.1 Discrete Optimal Control

This is a classic example of a problem in which gradient evaluations cost little more than function evaluations.

We begin with the continuous optimal control problems and discuss how gradients are computed and then move to the discretizations. We will not dwell on the functional analytic issues surrounding the rigorous definition of gradients of maps on function spaces, but the reader should be aware that control problems require careful attention to this. The most important results can be found in [151]. The function space setting for the particular control problems of interest in this section can be found in [170], [158], and [159], as can a discussion of more general problems.

The infinite-dimensional problem is

$$\min_u f, \tag{1.15}$$

where

$$f(u) = \int_0^T L(y(t), u(t), t)\, dt, \tag{1.16}$$

and we seek an optimal point $u \in L^\infty[0, T]$. u is called the *control variable* or simply the *control.* The function L is given and y, the *state variable*, satisfies the initial value problem (with $\dot{y} = dy/dt$)

$$\dot{y}(t) = \phi(y(t), u(t), t), y(0) = y_0. \tag{1.17}$$

One could view the problem (1.15)–(1.17) as a constrained optimization problem or, as we do here, think of the evaluation of f as requiring the solution of (1.17) before the integral on the right side of (1.16) can be evaluated. This means that evaluation of f requires the solution of (1.17), which is called the *state equation.*

$\nabla f(u)$, the gradient of f at u with respect to the L^2 inner product, is uniquely determined, if it exists, by

$$f(u+w) - f(u) - \int_0^T (\nabla f(u))(t) w(t)\, dt = o(\|w\|) \tag{1.18}$$

as $\|w\| \to 0$, uniformly in w. If $\nabla f(u)$ exists then

$$\int_0^T (\nabla f(u))(t) w(t)\, dt = \left.\frac{df(u+\xi w)}{d\xi}\right|_{\xi=0}.$$

If L and ϕ are continuously differentiable, then $\nabla f(u)$, as a function of t, is given by

$$\nabla f(u)(t) = p(t)\phi_u(y(t), u(t), t) + L_u(y(t), u(t), t). \tag{1.19}$$

In (1.19) p, the *adjoint variable*, satisfies the final-value problem on $[0, T]$

$$-\dot{p}(t) = p(t)\phi_y(y(t), u(t), t) + L_y(y(t), u(t), t), p(T) = 0. \tag{1.20}$$

So computing the gradient requires u and y, hence a solution of the state equation, and p, which requires a solution of (1.20), a final-value problem for the *adjoint equation.* In the general case, (1.17) is nonlinear, but (1.20) is a linear problem for p, which should be expected to be easier to solve. This is the motivation for our claim that a gradient evaluation costs little more than a function evaluation.

The discrete problems of interest here are constructed by solving (1.17) by numerical integration. After doing that, one can derive an adjoint variable and compute gradients using a discrete form of (1.19). However, in [139] the equation for the adjoint variable of the discrete problem is usually not a discretization of (1.20). For the forward Euler method, however, the discretization of the adjoint equation is the adjoint equation for the discrete problem and we use that discretization here for that reason.

The fully discrete problem is $\min_u f$, where $u \in R^N$ and

$$f(u) = \sum_{j=1}^N L((y)_j, (u)_j, j),$$

and the states $\{x_j\}$ are given by the Euler recursion

$$y_{j+1} = y_j + h\phi((y)_j, (u)_j, j) \text{ for } j = 0, \ldots, N-1,$$

where $h = T/(N-1)$ and x_0 is given. Then

$$(\nabla f(u))_j = (p)_j \phi_u((y)_j, (u)_j, j) + L_u((y)_j, (u)_j, j),$$

where $(p)_N = 0$ and

$$(p)_{j-1} = (p)_j + h\Big((p)_j \phi_y((y)_j, (u)_j, j) + L_y((y)_j, (u)_j, j)\Big) \text{ for } j = N, \ldots, 1.$$

1.6.2 Parameter Identification

This example, taken from [13], will appear throughout the book. The problem is small with $N = 2$. The goal is to identify the damping c and spring constant k of a linear spring by minimizing the difference of a numerical prediction and measured data. The experimental scenario is that the spring-mass system will be set into motion by an initial displacement from equilibrium and measurements of displacements will be taken at equally spaced increments in time.

The motion of an unforced harmonic oscillator satisfies the initial value problem

$$u'' + cu' + ku = 0; u(0) = u_0, u'(0) = 0, \tag{1.21}$$

on the interval $[0, T]$. We let $x = (c, k)^T$ be the vector of unknown parameters and, when the dependence on the parameters needs to be explicit, we will write $u(t : x)$ instead of $u(t)$ for the solution of (1.21). If the displacement is sampled at $\{t_j\}_{j=1}^M$, where $t_j = (j-1)T/(M-1)$, and the observations for u are $\{u_j\}_{j=1}^M$, then the objective function is

$$f(x) = \frac{1}{2}\sum_{j=1}^{M} |u(t_j : x) - u_j|^2. \tag{1.22}$$

This is an example of a *nonlinear least squares* problem.

u is differentiable with respect to x when $c^2 - 4k \neq 0$. In that case, the gradient of f is

$$\nabla f(x) = \begin{pmatrix} \sum_{j=1}^M \frac{\partial u(t_j:x)}{\partial c}(u(t_j : x) - u_j) \\ \sum_{j=1}^M \frac{\partial u(t_j:x)}{\partial k}(u(t_j : x) - u_j) \end{pmatrix}. \tag{1.23}$$

We can compute the derivatives of $u(t : x)$ with respect to the parameters by solving the *sensitivity equations*. Differentiating (1.21) with respect to c and k and setting $w_1 = \partial u/\partial c$ and $w_2 = \partial u/\partial k$ we obtain

$$\begin{aligned} &w_1'' + u' + cw_1' + kw_1 = 0; w_1(0) = w_1'(0) = 0, \\ &w_2'' + cw_2' + u + kw_2 = 0; w_2(0) = w_2'(0) = 0. \end{aligned} \tag{1.24}$$

If c is large, the initial value problems (1.21) and (1.24) will be *stiff* and one should use a good variable step stiff integrator. We refer the reader to [110], [8], [235] for details on this issue. In the numerical examples in this book we used the MATLAB code `ode15s` from [236]. Stiffness can also arise in the optimal control problem from §1.6.1 but does not in the specific examples we use in this book. We caution the reader that when one uses an ODE code the results may only be expected to be accurate to the tolerances input to the code. This limitation on the accuracy must be taken into account, for example, when approximating the Hessian by differences.

1.6.3 Convex Quadratics

While convex quadratic problems are, in a sense, the easiest of optimization problems, they present surprising challenges to the sampling algorithms presented in Part II and can illustrate fundamental problems with classical gradient-based methods like the steepest descent algorithm from §3.1. Our examples will all take $N = 2$, $b = 0$, and

$$H = \begin{pmatrix} \lambda_s & 0 \\ 0 & \lambda_l \end{pmatrix},$$

where $0 < \lambda_s \leq \lambda_l$. The function to be minimized is

$$f(x) = x^T H x$$

and the minimizer is $x^* = (0,0)^T$.

As λ_l/λ_s becomes large, the level curves of f become elongated. When $\lambda_s = \lambda_l = 1$, $\min_x f$ is the easiest problem in optimization.

1.7 Exercises on Basic Concepts

1.7.1. Prove Theorem 1.2.2.

1.7.2. Consider the parameter identification problem for $x = (c, k, \omega, \phi)^T \in R^4$ associated with the initial value problem

$$u'' + cu' + ku = \sin(\omega t + \phi); u(0) = 10, u'(0) = 0.$$

For what values of x is u differentiable? Derive the sensitivity equations for those values of x for which u is differentiable.

1.7.3. Solve the system of sensitivity equations from exercise 1.7.2 numerically for $c = 10$, $k = 1$, $\omega = \pi$, and $\phi = 0$ using the integrator of your choice. What happens if you use a nonstiff integrator?

1.7.4. Let $N = 2$, $d = (1,1)^T$, and let $f(x) = x^T d + x^T x$. Compute, by hand, the minimizer using conjugate gradient iteration.

1.7.5. For the same f as in exercise 1.7.4 solve the constrained optimization problem

$$\min_{x \in U} f(x),$$

where U is the circle centered at $(0,0)^T$ of radius $1/3$. You can solve this by inspection; no computer and very little mathematics is needed.

Chapter 2

Local Convergence of Newton's Method

By a local convergence method we mean one that requires that the initial iterate x_0 is close to a local minimizer x^* at which the sufficient conditions hold.

2.1 Types of Convergence

We begin with the standard taxonomy of convergence rates [84], [154], [211].

DEFINITION 2.1.1. *Let* $\{x_n\} \subset R^N$ *and* $x^* \in R^N$. *Then*

- $x_n \to x^*$ q-quadratically *if* $x_n \to x^*$ *and there is* $K > 0$ *such that*

$$\|x_{n+1} - x^*\| \le K\|x_n - x^*\|^2.$$

- $x_n \to x^*$ q-superlinearly with q-order $\alpha > 1$ *if* $x_n \to x^*$ *and there is* $K > 0$ *such that*

$$\|x_{n+1} - x^*\| \le K\|x_n - x^*\|^\alpha.$$

- $x_n \to x^*$ q-superlinearly *if*

$$\lim_{n\to\infty} \frac{\|x_{n+1} - x^*\|}{\|x_n - x^*\|} = 0.$$

- $x_n \to x^*$ q-linearly *with* q-factor $\sigma \in (0,1)$ *if*

$$\|x_{n+1} - x^*\| \le \sigma\|x_n - x^*\|$$

for n sufficiently large.

DEFINITION 2.1.2. *An iterative method for computing* x^* *is said to be* locally *(q-quadratically, q-superlinearly, q-linearly, etc.) convergent if the iterates converge to* x^* *(q-quadratically, q-superlinearly, q-linearly, etc.) given that the initial data for the iteration is sufficiently good.*

We remind the reader that a q-superlinearly convergent sequence is also q-linearly convergent with q-factor σ for any $\sigma > 0$. A q-quadratically convergent sequence is q-superlinearly convergent with q-order of 2.

In some cases the accuracy of the iteration can be improved by means that are external to the algorithm, say, by evaluation of the objective function and its gradient with increasing accuracy as the iteration progresses. In such cases, one has no guarantee that the accuracy of the iteration is monotonically increasing but only that the accuracy of the results is improving at a rate determined by the improving accuracy in the function–gradient evaluations. The concept of *r-type* convergence captures this effect.

DEFINITION 2.1.3. *Let* $\{x_n\} \subset R^N$ *and* $x^* \in R^N$. *Then* $\{x_n\}$ *converges to* x^* r-(quadratically, superlinearly, linearly) *if there is a sequence* $\{\xi_n\} \subset R$ *converging q-(quadratically, superlinearly, linearly) to* 0 *such that*

$$\|x_n - x^*\| \leq \xi_n.$$

We say that $\{x_n\}$ *converges* r-superlinearly *with* r-order $\alpha > 1$ *if* $\xi_n \to 0$ *q-superlinearly with q-order* α.

2.2 The Standard Assumptions

We will assume that local minimizers satisfy the *standard assumptions* which, like the standard assumptions for nonlinear equations in [154], will guarantee that Newton's method converges q-quadratically to x^*. We will assume throughout this book that f and x^* satisfy Assumption 2.2.1.

ASSUMPTION 2.2.1.

1. f *is twice differentiable and*

$$\|\nabla^2 f(x) - \nabla^2 f(y)\| \leq \gamma \|x - y\|. \tag{2.1}$$

2. $\nabla f(x^*) = 0$.

3. $\nabla^2 f(x^*)$ *is positive definite.*

We sometimes say that f is twice *Lipschitz continuously* differentiable with *Lipschitz constant* γ to mean that part 1 of the standard assumptions holds.

If the standard assumptions hold then Theorem 1.4.1 implies that x^* is a local minimizer of f. Moreover, the standard assumptions for nonlinear equations [154] hold for the system $\nabla f(x) = 0$. This means that all of the local convergence results for nonlinear equations can be applied to unconstrained optimization problems. In this chapter we will quote those results from nonlinear equations as they apply to unconstrained optimization. However, these statements must be understood in the context of optimization. We will use, for example, the fact that the Hessian (the Jacobian of ∇f) is positive definite at x^* in our solution of the linear equation for the Newton step. We will also use this in our interpretation of the Newton iteration.

2.3 Newton's Method

As in [154] we will define iterative methods in terms of the transition from a current iteration x_c to a new one x_+. In the case of a system of nonlinear equations, for example, x_+ is the root of the *local linear model* of F about x_c

$$M_c(x) = F(x_c) + F'(x_c)(x - x_c).$$

Solving $M_c(x_+) = 0$ leads to the standard formula for the Newton iteration

$$x_+ = x_c - F'(x_c)^{-1}F(x_c). \tag{2.2}$$

One could say that Newton's method for unconstrained optimization is simply the method for nonlinear equations applied to $\nabla f(x) = 0$. While this is technically correct if x_c is near a minimizer, it is utterly wrong if x_c is near a maximum. A more precise way of expressing the idea is to say that x_+ is a minimizer of the *local quadratic model* of f about x_c.

$$m_c(x) = f(x_c) + \nabla f(x_c)^T(x - x_c) + \frac{1}{2}(x - x_c)^T\nabla^2 f(x_c)(x - x_c).$$

If $\nabla^2 f(x_c)$ is positive definite, then the minimizer x_+ of m_c is the unique solution of $\nabla m_c(x) = 0$. Hence,

$$0 = \nabla m_c(x_+) = \nabla f(x_c) + \nabla^2 f(x_c)(x_+ - x_c).$$

Therefore,

$$x_+ = x_c - (\nabla^2 f(x_c))^{-1}\nabla f(x_c), \tag{2.3}$$

which is the same as (2.2) with F replaced by ∇f and F' by $\nabla^2 f$. Of course, x_+ is not computed by forming an inverse matrix. Rather, given x_c, $\nabla f(x_c)$ is computed and the linear equation

$$\nabla^2 f(x_c)s = -\nabla f(x_c) \tag{2.4}$$

is solved for the *step* s. Then (2.3) simply says that $x_+ = x_c + s$.

However, if u_c is far from a minimizer, $\nabla^2 f(u_c)$ could have negative eigenvalues and the quadratic model will not have local minimizers (see exercise 2.7.4), and M_c, the local linear model of ∇f about u_c, could have roots which correspond to local maxima or saddle points of m_c. Hence, we must take care when far from a minimizer in making a correspondence between Newton's method for minimization and Newton's method for nonlinear equations. In this chapter, however, we will assume that we are sufficiently near a local minimizer for the standard assumptions for local optimality to imply those for nonlinear equations (as applied to ∇f). Most of the proofs in this chapter are very similar to the corresponding results, [154], for nonlinear equations. We include them in the interest of completeness.

We begin with a lemma from [154], which we state without proof.

LEMMA 2.3.1. *Assume that the standard assumptions hold. Then there is $\delta > 0$ so that for all $x \in \mathcal{B}(\delta)$*

$$\|\nabla^2 f(x)\| \le 2\|\nabla^2 f(x^*)\|, \tag{2.5}$$

$$\|(\nabla^2 f(x))^{-1}\| \le 2\|(\nabla^2 f(x^*))^{-1}\|, \tag{2.6}$$

and

$$\|(\nabla^2 f(x^*))^{-1}\|^{-1}\|e\|/2 \le \|\nabla f(x)\| \le 2\|\nabla^2 f(x^*)\|\|e\|. \tag{2.7}$$

As a first example, we prove the local convergence for Newton's method.

THEOREM 2.3.2. *Let the standard assumptions hold. Then there are $K > 0$ and $\delta > 0$ such that if $x_c \in \mathcal{B}(\delta)$, the Newton iterate from x_c given by* (2.3) *satisfies*

$$\|e_+\| \le K\|e_c\|^2. \tag{2.8}$$

Proof. Let δ be small enough so that the conclusions of Lemma 2.3.1 hold. By Theorem 1.2.1

$$e_+ = e_c - \nabla^2 f(x_c)^{-1}\nabla f(x_c) = (\nabla^2 f(x_c))^{-1}\int_0^1 (\nabla^2 f(x_c) - \nabla^2 f(x^* + te_c))e_c\,dt.$$

By Lemma 2.3.1 and the Lipschitz continuity of $\nabla^2 f$,

$$\|e_+\| \leq (2\|(\nabla^2 f(x^*))^{-1}\|)\gamma\|e_c\|^2/2.$$

This completes the proof of (2.8) with $K = \gamma\|(\nabla^2 f(x^*))^{-1}\|$. □

As in the nonlinear equations setting, Theorem 2.3.2 implies that the complete iteration is locally quadratically convergent.

THEOREM 2.3.3. *Let the standard assumptions hold. Then there is $\delta > 0$ such that if $x_0 \in \mathcal{B}(\delta)$, the Newton iteration*

$$x_{n+1} = x_n - (\nabla^2 f(x_n))^{-1}\nabla f(x_n)$$

converges q-quadratically to x^*.

Proof. Let δ be small enough so that the conclusions of Theorem 2.3.2 hold. Reduce δ if needed so that $K\delta = \eta < 1$. Then if $n \geq 0$ and $x_n \in \mathcal{B}(\delta)$, Theorem 2.3.2 implies that

$$\|e_{n+1}\| \leq K\|e_n\|^2 \leq \eta\|e_n\| < \|e_n\| \tag{2.9}$$

and hence $x_{n+1} \in \mathcal{B}(\eta\delta) \subset \mathcal{B}(\delta)$. Therefore, if $x_n \in \mathcal{B}(\delta)$ we may continue the iteration. Since $x_0 \in \mathcal{B}(\delta)$ by assumption, the entire sequence $\{x_n\} \subset \mathcal{B}(\delta)$. (2.9) then implies that $x_n \to x^*$ q-quadratically. □

Newton's method, from the local convergence point of view, is exactly the same as that for nonlinear equations applied to the problem of finding a root of ∇f. We exploit the extra structure of positive definiteness of $\nabla^2 f$ with an implementation of Newton's method based on the *Cholesky factorization* [127], [249], [264]

$$\nabla^2 f(u) = LL^T, \tag{2.10}$$

where L is lower triangular and has a positive diagonal.

We terminate the iteration when ∇f is sufficiently small (see [154]). A natural criterion is to demand a relative decrease in $\|\nabla f\|$ and terminate when

$$\|\nabla f(x_n)\| \leq \tau_r\|\nabla f(x_0)\|, \tag{2.11}$$

where $\tau_r \in (0,1)$ is the desired reduction in the gradient norm. However, if $\|\nabla f(x_0)\|$ is very small, it may not be possible to satisfy (2.11) in floating point arithmetic and an algorithm based entirely on (2.11) might never terminate. A standard remedy is to augment the relative error criterion and terminate the iteration using a combination of relative and absolute measures of ∇f, i.e., when

$$\|\nabla f(x_n)\| \leq \tau_r\|\nabla f(x_0)\| + \tau_a. \tag{2.12}$$

In (2.12) τ_a is an absolute error tolerance. Hence, the termination criterion input to many of the algorithms presented in this book will be in the form of a vector $\tau = (\tau_r, \tau_a)$ of relative and absolute residuals.

ALGORITHM 2.3.1. $\texttt{newton}(x, f, \tau)$

1. $r_0 = \|\nabla f(x)\|$
2. *Do while* $\|\nabla f(x)\| > \tau_r r_0 + \tau_a$
   (a) *Compute* $\nabla^2 f(x)$
   (b) *Factor* $\nabla^2 f(x) = LL^T$
   (c) *Solve* $LL^T s = -\nabla f(x)$

(d) $x = x + s$

(e) *Compute* $\nabla f(x)$.

Algorithm newton, as formulated above, is not completely satisfactory. The value of the objective function f is never used and step 2b will fail if $\nabla^2 f$ is not positive definite. This failure, in fact, could serve as a signal that one is too far from a minimizer for Newton's method to be directly applicable. However, if we are near enough (see Exercise 2.7.8) to a local minimizer, as we assume in this chapter, all will be well and we may apply all the results from nonlinear equations.

2.3.1 Errors in Functions, Gradients, and Hessians

In the presence of errors in functions and gradients, however, the problem of unconstrained optimization becomes more difficult than that of root finding. We discuss this difference only briefly here and for the remainder of this chapter assume that gradients are computed exactly, or at least as accurately as f, say, either analytically or with automatic differentiation [129], [130]. However, we must carefully study the effects of errors in the evaluation of the Hessian just as we did those of errors in the Jacobian in [154].

A significant difference from the nonlinear equations case arises if only functions are available and gradients and Hessians must be computed with differences. A simple one-dimensional analysis will suffice to explain this. Assume that we can only compute f approximately. If we compute $\hat{f} = f + \epsilon_f$ rather than f, then a forward difference gradient with difference increment h

$$D_h f(x) = \frac{\hat{f}(x+h) - \hat{f}(x)}{h}$$

differs from f' by $O(h + \epsilon_f/h)$ and this error is minimized if $h = O(\sqrt{\epsilon_f})$. In that case the error in the gradient is $\epsilon_g = O(h) = O(\sqrt{\epsilon_f})$. If a forward difference Hessian is computed using D_h as an approximation to the gradient, then the error in the Hessian will be

$$\Delta = O(\sqrt{\epsilon_g}) = O(\epsilon_f^{1/4}) \tag{2.13}$$

and the accuracy in $\nabla^2 f$ will be much less than that of a Jacobian in the nonlinear equations case.

If ϵ_f is significantly larger than machine roundoff, (2.13) indicates that using numerical Hessians based on a second numerical differentiation of the objective function will not be very accurate. Even in the best possible case, where ϵ_f is the same size as machine roundoff, finite difference Hessians will not be very accurate and will be very expensive to compute if the Hessian is dense. If, as on most computers today, machine roundoff is (roughly) 10^{-16}, (2.13) indicates that a forward difference Hessian will be accurate to roughly four decimal digits.

One can obtain better results with centered differences, but at a cost of twice the number of function evaluations. A centered difference approximation to ∇f is

$$D_h f(x) = \frac{\hat{f}(x+h) - \hat{f}(x-h)}{2h}$$

and the error is $O(h^2 + \epsilon_f/h)$, which is minimized if $h = O(\epsilon_f^{1/3})$ leading to an error in the gradient of $\epsilon_g = O(\epsilon_f^{2/3})$. Therefore, a central difference Hessian will have an error of

$$\Delta = O((\epsilon_g)^{2/3}) = O(\epsilon_f^{4/9}),$$

which is substantially better. We will find that accurate gradients are much more important than accurate Hessians and one option is to compute gradients with central differences and Hessians with forward differences. If one does that the centered difference gradient error is $O(\epsilon_f^{2/3})$ and therefore the forward difference Hessian error will be

$$\Delta = O\left(\sqrt{\epsilon_g}\right) = O(\epsilon_f^{1/3}).$$

More elaborate schemes [22] compute a difference gradient and then reuse the function evaluations in the Hessian computation.

In many optimization problems, however, accurate gradients are available. When that is the case, numerical differentiation to compute Hessians is, like numerical computation of Jacobians for nonlinear equations [154], a reasonable idea for many problems and the less expensive forward differences work well.

Clever implementations of difference computation can exploit sparsity in the Hessian [67], [59] to evaluate a forward difference approximation with far fewer than N evaluations of ∇f. In the sparse case it is also possible [22], [23] to reuse the points from a centered difference approximation to the gradient to create a second-order accurate Hessian.

Unless $\epsilon_g(x_n) \rightarrow 0$ as the iteration progresses, one cannot expect convergence. For this reason estimates like (2.14) are sometimes called *local improvement* [88] results. Theorem 2.3.4 is a typical example.

THEOREM 2.3.4. *Let the standard assumptions hold. Then there are* $\bar{K} > 0$, $\delta > 0$, *and* $\delta_1 > 0$ *such that if* $x_c \in \mathcal{B}(\delta)$ *and* $\|\Delta(x_c)\| < \delta_1$ *then*

$$x_+ = x_c - (\nabla^2 f(x_c) + \Delta(x_c))^{-1}(\nabla f(x_c) + \epsilon_g(x_c))$$

is defined (i.e., $\nabla^2 f(x_c) + \Delta(x_c)$ *is nonsingular) and satisfies*

$$\|e_+\| \le \bar{K}(\|e_c\|^2 + \|\Delta(x_c)\|\|e_c\| + \|\epsilon_g(x_c)\|). \tag{2.14}$$

Proof. Let δ be small enough so that the conclusions of Lemma 2.3.1 and Theorem 2.3.2 hold. Let

$$x_+^N = x_c - (\nabla^2 f(x_c))^{-1}\nabla f(x_c)$$

and note that

$$x_+ = x_+^N + ((\nabla^2 f(x_c))^{-1} - (\nabla^2 f(x_c) + \Delta(x_c))^{-1})\nabla f(x_c) - (\nabla^2 f(x_c) + \Delta(x_c))^{-1}\epsilon_g(x_c).$$

Lemma 2.3.1 and Theorem 2.3.2 imply

$$\begin{aligned} \|e_+\| \le \; & K\|e_c\|^2 + 2\|(\nabla^2 f(x_c))^{-1} - (\nabla^2 f(x_c) + \Delta(x_c))^{-1}\|\|\nabla^2 f(x^*)\|\|e_c\| \\ & + \|(\nabla^2 f(x_c) + \Delta(x_c))^{-1}\|\|\epsilon_g(x_c)\|. \end{aligned} \tag{2.15}$$

If

$$\|\Delta(x_c)\| \le \|(\nabla^2 f(x^*))^{-1}\|^{-1}/4,$$

then Lemma 2.3.1 implies that

$$\|\Delta(x_c)\| \le \|(\nabla^2 f(x_c))^{-1}\|^{-1}/2$$

and the Banach Lemma [12], [154] states that $\nabla^2 f(x_c) + \Delta(x_c)$ is nonsingular and

$$\|(\nabla^2 f(x_c) + \Delta(x_c))^{-1}\| \le 2\|(\nabla^2 f(x_c))^{-1}\| \le 4\|(\nabla^2 f(x^*))^{-1}\|.$$

Hence,

$$\|(\nabla^2 f(x_c))^{-1} - (\nabla^2 f(x_c) + \Delta(x_c))^{-1}\| \le 8\|(\nabla^2 f(x^*))^{-1}\|^2 \|\Delta(x_c)\|.$$

We use these estimates and (2.15) to obtain

$$\|e_+\| \le K\|e_c\|^2 + 16\|(\nabla^2 f(x^*))^{-1}\|^2 \|\nabla^2 f(x^*)\| \|\Delta(x_c)\| \|e_c\| + 4\|(\nabla^2 f(x^*))^{-1}\| \|\epsilon_g(x_c)\|.$$

Setting

$$\bar{K} = K + 16\|(\nabla^2 f(x^*))^{-1}\|^2 \|\nabla^2 f(x^*)\| + 4\|(\nabla^2 f(x^*))^{-1}\|$$

completes the proof. □

As is the case with equations, (2.14) implies that one cannot hope to find a minimizer with more accuracy that one can evaluate ∇f. In most cases the iteration will *stagnate* once $\|e\|$ is (roughly) the same size as ϵ_g. The speed of convergence will be governed by the accuracy in the Hessian.

The result for the *chord method* illustrates this latter point. In the chord method we form and compute the Cholesky factorization of $\nabla^2 f(x_0)$ and use that factorization to compute all subsequent Newton steps. Hence,

$$x_+ = x_c - (\nabla^2 f(x_0))^{-1} \nabla f(x_c)$$

and

$$\|\Delta(x_c)\| \le \gamma\|x_0 - x_c\| \le \gamma(\|e_0\| + \|e_c\|). \tag{2.16}$$

Algorithmically the chord iteration differs from the Newton iteration only in that the computation and factorization of the Hessian is moved outside of the main loop.

ALGORITHM 2.3.2. $\texttt{chord}(x, f, \tau)$

1. $r_0 = \|\nabla f(x)\|$
2. *Compute* $\nabla^2 f(x)$
3. *Factor* $\nabla^2 f(x) = LL^T$
4. *Do while* $\|\nabla f(x)\| > \tau_r r_0 + \tau_a$
   (a) *Solve* $LL^T s = -\nabla f(x)$
   (b) $x = x + s$
   (c) *Compute* $\nabla f(x)$.

The convergence theory follows from Theorem 2.3.4 with $\epsilon_g = 0$ and $\Delta = O(\|e_0\|)$.

THEOREM 2.3.5. *Let the standard assumptions hold. Then there are* $K_C > 0$ *and* $\delta > 0$ *such that if* $x_0 \in \mathcal{B}(\delta)$ *the chord iterates converge q-linearly to* x^* *and*

$$\|e_{n+1}\| \le K_C \|e_0\| \|e_n\|. \tag{2.17}$$

Proof. Let δ be small enough so that the conclusions of Theorem 2.3.4 hold. Assume that $e_n, e_0 \in \mathcal{B}(\delta)$. Combining (2.16) and (2.14) implies

$$\|e_{n+1}\| \le \bar{K}(\|e_n\|(1+\gamma) + \gamma\|e_0\|)\|e_n\| \le \bar{K}(1+2\gamma)\delta\|e_n\|.$$

Hence, if δ is small enough so that

$$\bar{K}(1 + 2\gamma)\delta = \eta < 1,$$

then the chord iterates converge q-linearly to x^*. Q-linear convergence implies that $\|e_n\| < \|e_0\|$ and hence (2.17) holds with $K_C = \bar{K}(1 + 2\gamma)$. $\square$

The Shamanskii method [233], [154], [211] is a generalization of the chord method that updates Hessians after every $m + 1$ nonlinear iterations. Newton's method corresponds to $m = 0$ and the chord method to $m = \infty$. The convergence result is a direct consequence of Theorems 2.3.3 and 2.3.5.

THEOREM 2.3.6. *Let the standard assumptions hold and let $m \geq 1$ be given. Then there are $K_S > 0$ and $\delta > 0$ such that if $x_0 \in \mathcal{B}(\delta)$ the Shamanskii iterates converge q-superlinearly to x^* with q-order $m + 1$ and*

$$\|e_{n+1}\| \leq K_S \|e_n\|^{m+1}. \tag{2.18}$$

As one more application of Theorem 2.3.4, we analyze the effects of a difference approximation of the Hessian. We follow the notation of [154] where possible. For example, to construct a Hessian matrix, whose columns are $\nabla^2 f(x) e_j$, where e_j is the unit vector with jth component 1 and other components 0, we could approximate the matrix–vector products $\nabla^2 f(x) e_j$ by forward differences and then symmetrize the resulting matrix. We define

$$\nabla_h^2 f(x) = (A_h + A_h^T)/2, \tag{2.19}$$

where A_h is the matrix whose jth column is $D_h^2 f(x : e_j)$. $D_h^2 f(x : w)$, the difference approximation of the action of the Hessian $\nabla^2 f(x)$ on a vector w, is defined to be the quotient

$$D_h^2 f(x : w) = \begin{cases} 0, & w = 0, \\ \dfrac{\nabla f(x + hw/\|w\|) - \nabla f(x)}{h/\|w\|}, & w \neq 0. \end{cases} \tag{2.20}$$

Note that we may also write

$$D_h^2 f(x : w) = D_h(\nabla f)(x : w),$$

where the notation D_h, taken from [154], denotes numerical directional derivative. If $\|x\|$ is very large, then the error in computing the sum $x + hw/\|w\|$ will have to be taken into consideration in the choice of h.

We warn the reader, as we did in [154], that $D^2 f(x : w)$ is not a linear map and that $D^2 f(x : w)$ is not, in general, the same as $\nabla_h^2 f(x) w$.

If we compute $\nabla f(x) + \epsilon_g(x)$ and the gradient errors satisfy an estimate of the form

$$\|\epsilon_g(x)\| \leq \bar{\epsilon}$$

for all x, then the computed difference Hessian is $\nabla_h(\nabla f + \epsilon_g)$ and satisfies

$$\|\nabla^2 f(x) - \nabla_h(\nabla f + \epsilon_g)(x)\| = O(h + \bar{\epsilon}/h). \tag{2.21}$$

So, as in [154], the choice $h \approx \sqrt{\bar{\epsilon}}$ is optimal in the sense that it minimizes the quantity in the O-term in (2.21).

The local convergence theorem in this case is [88], [154], [278], as follows.

THEOREM 2.3.7. *Let the standard assumptions hold. Then there are* δ, $\bar{\epsilon}$, *and* $K_D > 0$ *such that if* $x_c \in \mathcal{B}(\delta)$, $\|\epsilon_g(x)\| \leq \bar{\epsilon}_c$ *for all* $x \in \mathcal{B}(\delta)$, *and*

$$h \geq M\sqrt{\|\epsilon_g(x_c)\|}$$

then

$$x_+ = x_c - (\nabla_{h_c}(\nabla f(x_c) + \epsilon_g(x_c)))^{-1}(\nabla f(x_c) + \epsilon_g(x_c))$$

satisfies

$$\|e_+\| \leq K_D(\bar{\epsilon}_c + (\bar{\epsilon}_c + h)\|e_c\|).$$

2.3.2 Termination of the Iteration

It is not safe to terminate the iteration when $f(x_c) - f(x_+)$ is small, and no conclusions can safely be drawn by examination of the differences of the objective function values at successive iterations. While some of the algorithms for difficult problems in Part II of this book do indeed terminate when successive function values are close, this is an act of desperation. For example, if

$$f(x_n) = -\sum_{j=1}^{n} j^{-1},$$

then $f(x_n) \to -\infty$ but $f(x_{n+1}) - f(x_n) = -1/(n+1) \to 0$. The reader has been warned.

If the standard assumptions hold, then one may terminate the iteration when the norm of ∇f is sufficiently small relative to $\nabla f(x_0)$ (see [154]). We will summarize the key points here and refer the reader to [154] for the details. The idea is that if $\nabla^2 f(x^*)$ is well conditioned, then a small gradient norm implies a small error norm. Hence, for any gradient-based iterative method, termination on small gradients is reasonable. In the special case of Newton's method, the norm of the step is a very good indicator of the error and if one is willing to incur the added cost of an extra iteration, a very sharp bound on the error can be obtained, as we will see below.

LEMMA 2.3.8. *Assume that the standard assumptions hold. Let* $\delta > 0$ *be small enough so that the conclusions of Lemma* 2.3.1 *hold for* $x \in \mathcal{B}(\delta)$. *Then for all* $x \in \mathcal{B}(\delta)$

$$\frac{\|e\|}{4\|e_0\|\kappa(\nabla^2 f(x^*))} \leq \frac{\|\nabla f(x)\|}{\|\nabla f(x_0)\|} \leq \frac{4\kappa(\nabla^2 f(x^*))\|e\|}{\|e_0\|}. \tag{2.22}$$

The meaning of (2.22) is that, up to a constant multiplier, the norm of the relative gradient is the same as the norm of the relative error. This partially motivates the termination condition (2.12).

In the special case of Newton's method, one can use the steplength as an accurate estimate of the error because Theorem 2.3.2 implies that

$$\|e_c\| = \|s\| + O(\|e_c\|^2). \tag{2.23}$$

Hence, near the solution s and e_c are essentially the same size. The cost of using (2.23) is that all the information needed to compute $x_+ = x_c + s$ has been computed. If we terminate the iteration when $\|s\|$ is smaller than our desired tolerance and then take x_+ as the final result, we have attained more accuracy than we asked for. One possibility is to terminate the iteration when $\|s\| = O(\sqrt{\tau_s})$ for some $\tau_s > 0$. This, together with (2.23), will imply that $\|e_c\| = O(\sqrt{\tau_s})$ and hence, using Theorem 2.3.2, that

$$\|e_+\| = O(\|e_c\|^2) = O(\tau_s). \tag{2.24}$$

For a superlinearly convergent method, termination on small steps is equally valid but one cannot use (2.24). For a superlinearly convergent method we have

$$\|e_c\| = \|s\| + o(\|e_c\|) \text{ and } \|e_+\| = o(\|e_c\|). \tag{2.25}$$

Hence, we can conclude that $\|e_+\| < \tau_s$ if $\|s\| < \tau_s$. This is a weaker, but still very useful, result.

For a q-linearly convergent method, such as the chord method, making termination decisions based on the norms of the steps is much riskier. The relative error in estimating $\|e_c\|$ by $\|s\|$ is

$$\frac{|\|e_c\| - \|s\||}{\|e_c\|} \le \frac{\|e_c + s\|}{\|e_c\|} = \frac{\|e_+\|}{\|e_c\|}.$$

Hence, estimation of errors by steps is worthwhile only if convergence is fast. One can go further [156] if one has an estimate ρ of the q-factor that satisfies

$$\|e_+\| \le \rho\|e_c\|.$$

In that case,

$$(1-\rho)\|e_c\| \le \|e_c\| - \|e_+\| \le \|e_c - e_+\| = \|s\|.$$

Hence

$$\|e_+\| \le \rho\|e_c\| \le \frac{\rho}{1-\rho}\|s\|. \tag{2.26}$$

So, if the q-factor can be estimated from above by ρ and

$$\|s\| < (1-\rho)\tau_s/\rho,$$

then $\|e_+\| < \tau_s$. This approach is used in ODE and DAE codes [32], [234], [228], [213], but requires good estimates of the q-factor and we do not advocate it for q-linearly convergent methods for optimization. The danger is that if the convergence is slow, the approximate q-factor can be a gross underestimate and cause premature termination of the iteration.

It is not uncommon for evaluations of f and ∇f to be very expensive and optimizations are, therefore, usually allocated a fixed maximum number of iterations. Some algorithms, such as the DIRECT, [150], algorithm we discuss in §8.4.2, assign a limit to the number of function evaluations and terminate the iteration in only this way.

2.4 Nonlinear Least Squares

Nonlinear least squares problems have objective functions of the form

$$f(x) = \frac{1}{2}\sum_{i=1}^{M} \|r_i(x)\|_2^2 = \frac{1}{2}R(x)^T R(x). \tag{2.27}$$

The vector $R = (r_1, \dots, r_M)$ is called the *residual*. These problems arise in data fitting, for example. In that case M is the number of observations and N is the number of parameters; for these problems $M > N$ and we say the problem is *overdetermined*. If $M = N$ we have a nonlinear equation and the theory and methods from [154] are applicable. If $M < N$ the problem is *underdetermined*. Overdetermined least squares problems arise most often in data fitting applications like the parameter identification example in §1.6.2. Underdetermined problems are less common, but are, for example, important in the solution of high-index differential algebraic equations [48], [50].

The local convergence theory for underdetermined problems has the additional complexity that the limit of the Gauss–Newton iteration is not uniquely determined by the distance of the initial iterate to the set of points where $R(x^*) = 0$. In §2.4.3 we describe the difficulties and state a simple convergence result.

If x^* is a local minimizer of f and $f(x^*) = 0$, the problem $\min f$ is called a *zero residual problem* (a remarkable and suspicious event in the data fitting scenario). If $f(x^*)$ is small, the expected result in data fitting if the model (i.e., R) is good, the problem is called a *small residual problem*. Otherwise one has a *large residual problem*.

Nonlinear least squares problems are an intermediate stage between nonlinear equations and optimization problems and the methods for their solution reflect this. We define the $M \times N$ Jacobian R' of R by

$$(R'(x))_{ij} = \partial r_i/\partial x_j,\ 1 \le i \le M,\ 1 \le j \le N. \tag{2.28}$$

With this notation it is easy to show that

$$\nabla f(x) = R'(x)^T R(x) \in R^N. \tag{2.29}$$

The necessary conditions for optimality imply that at a minimizer x^*

$$R'(x^*)^T R(x^*) = 0. \tag{2.30}$$

In the underdetermined case, if $R'(x^*)$ has full row rank, (2.30) implies that $R(x^*) = 0$; this is not the case for overdetermined problems.

The cost of a gradient is roughly that of a Jacobian evaluation, which, as is the case with nonlinear equations, is the most one is willing to accept. Computation of the Hessian (an $N \times N$ matrix)

$$\nabla^2 f(x) = R'(x)^T R'(x) + \sum_{j=1}^{M} r_i(x)^T \nabla^2 r_i(x) \tag{2.31}$$

requires computation of the M Hessians $\{\nabla^2 r_i\}$ for the second-order term

$$\sum_{j=1}^{M} r_i(x)^T \nabla^2 r_i(x)$$

and is too costly to be practical.

We will also express the second-order term as

$$\sum_{j=1}^{M} r_i(x)^T \nabla^2 r_i(x) = R''(x)^T R(x),$$

where the second derivative R'' is a tensor. The notation is to be interpreted in the following way. For $v \in R^M$, $R''(x)^T v$ is the $N \times N$ matrix

$$R''(x)^T v = \nabla^2 (R(x)^T v) = \sum_{i=1}^{M} (v)_i \nabla^2 r_i(x).$$

We will use the tensor notation when expanding R about x^* in some of the analysis to follow.

2.4.1 Gauss–Newton Iteration

The Gauss–Newton algorithm simply discards the second-order term in $\nabla^2 f$ and computes a step

$$\begin{aligned} s &= -(R'(x_c)^T R'(x_c))^{-1} \nabla f(x_c) \\ &= -(R'(x_c)^T R'(x_c))^{-1} R'(x_c)^T R(x_c). \end{aligned} \tag{2.32}$$

The Gauss–Newton iterate is $x_+ = x_c + s$. One motivation for this approach is that $R''(x)^T R(x)$ vanishes for zero residual problems and therefore might be negligible for small residual problems.

Implicit in (2.32) is the assumption that $R'(x_c)^T R'(x_c)$ is nonsingular, which implies that $M \geq N$. Another interpretation, which also covers underdetermined problems, is to say that the Gauss–Newton iterate is the *minimum norm* solution of the local linear model of our nonlinear least squares problem

$$\min \frac{1}{2}\|R(x_c) + R'(x_c)(x - x_c)\|^2. \tag{2.33}$$

Using (2.33) and linear least squares methods is a more accurate way to compute the step than using (2.32), [115], [116], [127]. In the underdetermined case, the Gauss–Newton step can be computed with the singular value decomposition [49], [127], [249]. (2.33) is an overdetermined, square, or underdetermined linear least squares problem if the nonlinear problem is overdetermined, square, or underdetermined.

The standard assumptions for nonlinear least squares problems follow in Assumption 2.4.1.

ASSUMPTION 2.4.1. *x^* is a minimizer of $\|R\|_2^2$, R is Lipschitz continuously differentiable near x^*, and $R'(x^*)^T R'(x^*)$ has maximal rank. The rank assumption may also be stated as*

- *$R'(x^*)$ is nonsingular ($M = N$),*
- *$R'(x^*)$ has full column rank ($M > N$),*
- *$R'(x^*)$ has full row rank ($M < N$).*

2.4.2 Overdetermined Problems

THEOREM 2.4.1. *Let $M > N$. Let Assumption 2.4.1 hold. Then there are $K > 0$ and $\delta > 0$ such that if $x_c \in \mathcal{B}(\delta)$ then the error in the Gauss–Newton iteration satisfies*

$$\|e_+\| \leq K(\|e_c\|^2 + \|R(x^*)\|\|e_c\|). \tag{2.34}$$

Proof. Let δ be small enough so that $\|x - x^*\| < \delta$ implies that $R'(x)^T R'(x)$ is nonsingular. Let γ be the Lipschitz constant for R'.

By (2.32)

$$\begin{aligned} e_+ &= e_c - (R'(x_c)^T R'(x_c))^{-1} R'(x_c)^T R(x_c) \\ &= (R'(x_c)^T R'(x_c))^{-1} R'(x_c)^T (R'(x_c)e_c - R(x_c)). \end{aligned} \tag{2.35}$$

Note that

$$\begin{aligned} R'(x_c)e_c - R(x_c) &= R'(x_c)e_c - R(x^*) + R(x^*) - R(x_c) \\ &= -R(x^*) + (R'(x_c)e_c + R(x^*) - R(x_c)). \end{aligned}$$

Now,

$$\|R'(x_c)e_c + R(x^*) - R(x_c)\| \leq \gamma\|e_c\|^2/2$$

and, since $R'(x^*)^T R(x^*) = 0$,

$$-R'(x_c)^T R(x^*) = (R'(x^*) - R'(x_c))^T R(x^*).$$

Hence,

$$\begin{aligned} \|e_+\| &\le \|(R'(x_c)^T R'(x_c))^{-1}\| \|(R'(x^*) - R'(x_c))^T R(x^*)\| \\ &+ \frac{\|(R'(x_c)^T R'(x_c))^{-1}\| \|R'(x_c)^T\| \gamma \|e_c\|^2}{2} \\ &\le \|(R'(x_c)^T R'(x_c))^{-1}\| \gamma \|e_c\| \left(\frac{\|R(x^*)\| + \|R'(x_c)^T\| \|e_c\|}{2} \right). \end{aligned} \tag{2.36}$$

Setting

$$K = \gamma \max_{x \in \mathcal{B}(\delta)} (\|(R'(x)^T R'(x))^{-1}\| \left(\frac{1 + \|R'(x)^T\|}{2} \right)$$

completes the proof. □

There are several important consequences of Theorem 2.4.1. The first is that for zero residual problems, the local convergence rate is q-quadratic because the $\|R(x^*)\|\|e_c\|$ term on the right side of (2.34) vanishes. For a problem other than a zero residual one, even q-linear convergence is not guaranteed. In fact, if $x_c \in \mathcal{B}(\delta)$ then (2.35) will imply that $\|e_+\| \le r\|e_c\|$ for some $0 < r < 1$ if

$$K(\delta + \|R'(x^*)\|) \le r \tag{2.37}$$

and therefore the q-factor will be $K\|R'(x^*)\|$. Hence, for small residual problems and accurate initial data the convergence of Gauss–Newton will be fast. Gauss–Newton may not converge at all for large residual problems.

Equation (2.36) exposes a more subtle issue when the term

$$(R'(x^*) - R'(x_c))^T R(x^*)$$

is considered as a whole, rather than estimated by

$$\gamma \|e_c\| \|R(x^*)\|.$$

Using Taylor's theorem and the necessary conditions ($R'(x^*)^T R(x^*) = 0$) we have

$$\begin{aligned} R'(x_c)^T R(x^*) &= [R'(x^*) + R''(x^*) e_c + O(\|e_c\|^2)]^T R(x^*) \\ &= e_c^T R''(x^*)^T R(x^*) + O(\|e_c\|^2). \end{aligned}$$

Recall that

$$R''(x^*)^T R(x^*) = \nabla^2 f(x^*) - R'(x^*)^T R'(x^*)$$

and hence

$$\begin{aligned} &\|(R'(x^*) - R'(x_c))^T R(x^*)\| \\ &\le \|\nabla^2 f(x^*) - R'(x^*)^T R'(x^*)\| \|R(x^*)\| + O(\|e_c\|^2). \end{aligned} \tag{2.38}$$

In a sense (2.38) says that even for a large residual problem, convergence can be fast if the problem is not very nonlinear (small R''). In the special case of a linear least squares problem (where $R'' = 0$) Gauss–Newton becomes simply the solution of the normal equations and converges in one iteration.

So, Gauss–Newton can be expected to work well for overdetermined small residual problems and good initial iterates. For large residual problems and/or initial data far from the solution, there is no reason to expect Gauss–Newton to give good results. We address these issues in §3.2.3.

2.4.3 Underdetermined Problems

We begin with the linear underdetermined least squares problem

$$\min \|Ax - b\|^2. \tag{2.39}$$

If A is $M \times N$ with $M < N$ there will not be a unique minimizer but there will be a unique minimizer with minimum norm. The *minimum norm* solution can be expressed in terms of the *singular value decomposition* [127], [249] of A,

$$A = U\Sigma V^T. \tag{2.40}$$

In (2.40), Σ is an $N \times N$ diagonal matrix. The diagonal entries of Σ, $\{\sigma_i\}$ are called the *singular values.* $\sigma_i \geq 0$ and $\sigma_i = 0$ if $i > M$. The columns of the $M \times N$ matrix U and the $N \times N$ matrix V are called the left and right *singular vectors.* U and V have orthonormal columns and hence the minimum norm solution of (2.39) is

$$x = A^\dagger b,$$

where $A^\dagger = V\Sigma^\dagger U^T$,

$$\Sigma^\dagger = \text{diag}(\sigma_1^\dagger, \dots, \sigma_N^\dagger),$$

and

$$\sigma_i^\dagger = \begin{cases} \sigma_i^{-1}, & \sigma_i \neq 0, \\ 0, & \sigma_i = 0. \end{cases}$$

$A^\dagger$ is called the *Moore–Penrose inverse* [49], [189], [212]. If A is a square nonsingular matrix, then $A^\dagger = A^{-1}$; if $M > N$ then the definition of $A^\dagger$ using the singular value decomposition is still valid; and, if A has full column rank, $A^\dagger = (A^T A)^{-1} A^T$.

Two simple properties of the Moore–Penrose inverse are that $A^\dagger A$ is a projection onto the range of $A^\dagger$ and $AA^\dagger$ is a projection onto the range of A. This means that

$$A^\dagger A A^\dagger = A^\dagger, (A^\dagger A)^T = A^\dagger A, AA^\dagger A = A, \text{ and } (AA^\dagger)^T = AA^\dagger. \tag{2.41}$$

So the minimum norm solution of the local linear model (2.33) of an underdetermined nonlinear least squares problem can be written as [17], [102]

$$s = -R'(x_c)^\dagger R(x_c) \tag{2.42}$$

and the Gauss–Newton iteration [17] is

$$x_+ = x_c - R'(x_c)^\dagger R(x_c). \tag{2.43}$$

The challenge in formulating a local convergence result is that there is not a unique optimal point that attracts the iterates.

In the linear case, where $R(x) = Ax - b$, one gets

$$x_+ = x_c - A^\dagger(Ax_c - b) = (I - A^\dagger A)x_c - A^\dagger b.$$

Set $e = x - A^\dagger b$ and note that

$$A^\dagger A A^\dagger b = A^\dagger b$$

by (2.41). Hence

$$e_+ = (I - A^\dagger A)e_c.$$

This does not imply that $x_+ = A^\dagger b$, the minimum norm solution, only that x_+ is a solution of the problem and the iteration converges in one step. The Gauss–Newton iteration cannot correct for errors that are not in the range of $A^\dagger$.

Let

$$\mathcal{Z} = \{x \mid R(x) = 0\}.$$

We show in Theorem 2.4.2, a special case of the result in [92], that if the standard assumptions hold at a point $x^* \in \mathcal{Z}$, then the iteration will converge q-quadratically to a point $z^* \in Z$. However, there is no reason to expect that $z^* = x^*$. In general z^* will depend on x_0, a very different situation from the overdetermined case. The hypotheses of Theorem 2.4.2, especially that of full column rank in $R'(x)$, are less general than those in [24], [17], [25], [92], and [90].

THEOREM 2.4.2. *Let $M \leq N$ and let Assumption* 2.4.1 *hold for some $x^* \in \mathcal{Z}$. Then there is $\delta > 0$ such that if*

$$\|x_0 - x^*\| \leq \delta,$$

then the Gauss–Newton iterates

$$x_{n+1} = x_n - R'(x_n)^\dagger R(x_n)$$

exist and converge r-quadratically to a point $z^* \in \mathcal{Z}$.

Proof. Assumption 2.4.1 and results in [49], [126] imply that if δ is sufficiently small then there is ρ_1 such that $R'(x)^\dagger$ is Lipschitz continuous in the set

$$\mathcal{B}_1 = \{x \mid \|x - x^*\| \leq \rho_1\}$$

and the singular values of $R'(x)$ are bounded away from zero in $\mathcal{B}_1$. We may, reducing ρ_1 if necessary, apply the Kantorovich theorem [154], [151], [211] to show that if $x \in \mathcal{B}_1$ and $w \in \mathcal{Z}$ is such that

$$\|x - w\| = \min_{z \in \mathcal{Z}} \|x - z\|,$$

then there is $\xi = \xi(x) \in Z$ such that

$$\|w - \xi(x)\| = O(\|x - w\|^2) \leq \|x - w\|/2$$

and ξ is in the range of $R'(w)^\dagger$, i.e.,

$$R'(w)^\dagger R'(w)(x - \xi(x)) = x - \xi(x).$$

The method of the proof is to adjust δ so that the Gauss–Newton iterates remain in $\mathcal{B}_1$ and $R(x_n) \to 0$ sufficiently rapidly. We begin by requiring that $\delta < \rho_1/2$.

Let $x_c \in \mathcal{B}_1$ and let $e = x - \xi(x_c)$. Taylor's theorem, the fundamental theorem of calculus, and (2.41) imply that

$$\begin{aligned} e_+ &= e_c - R'(x_c)^\dagger R(x_c) \\ &= e_c - (R'(x_c)^\dagger - R'(\xi(x))^\dagger)R(x) - R'(x*)^\dagger R(x) \\ &= e_0 - R'(x^*)^\dagger R(x) + O(\|e_c\|^2) \\ &= (I - R'(x^*)^\dagger R'(x^*))e_c + O(\|e_c\|^2) = O(\|e_c\|^2). \end{aligned}$$

If we define $d(x) = \min_{z \in \mathcal{Z}} \|x - z\|$ then there is K_1 such that

$$d(x_+) \leq \|x_+ - \xi(x_c)\| \leq K_1\|x_c - \xi(x_c)\|^2 \leq K_1 d(x_c)^2. \tag{2.44}$$

Now let

$$\rho_2 = \min(\rho_1, (2K_1)^{-1}).$$

So if

$$x_c \in \mathcal{B}_2 = \{x \mid \|x - x^*\| \le \rho_2\}$$

then

$$d(x_+) \le d(x_c)/2. \tag{2.45}$$

Finally, there is K_2 such that

$$\begin{aligned} \|x_+ - x_*\| &\le \|x_c - x^*\| + \|x_+ - x_c\| = \|x_c - x^*\| + \|R'(x_c)^\dagger R(x_c)\| \\ &\le \|x_c - x^*\| + K_2\|x_c - \xi(x_c)\|. \end{aligned}$$

We now require that

$$\delta \le \frac{\rho_2}{2(1 + K_2)}. \tag{2.46}$$

We complete the proof by induction. If $\|x_0 - x^*\| \le \delta$ and the Gauss–Newton iterates $\{x_k\}_{k=0}^n$ are in $\mathcal{B}_2$, then x_{n+1} is be defined and, using (2.46) and (2.44),

$$\|x_{n+1} - x^*\| \le \|x_0 - x^*\| + K_3 \sum_{k=0}^{n+1} d(x_k) \le \delta + 2K_3 d(x_0) \le \rho_1.$$

Hence, the Gauss–Newton iterates exist, remain in $\mathcal{B}_0$, and $d_n \to 0$.

To show that the sequence of Gauss–Newton iterates does in fact converge, we observe that there is K_3 such that

$$\|x_+ - x_c\| = \|R'(x_c)^\dagger R(x_c)\| \le K_3\|x_c - \xi(x_c)\| \le K_3 d(x_c).$$

Therefore (2.45) implies that for any $m, n \ge 0$,

$$\begin{aligned} \|x_{n+m} - x_n\| &\le \textstyle\sum_{l=n}^{n+m-1} \|x_{l+1} - x_l\| \\ &= \textstyle\sum_{l=n}^{n+m} d(x_l) = d(x_n)\dfrac{1 - 2^{-m}}{2} \\ &\le 2d(x_n) \le 2^{-n+1} d(x_0). \end{aligned}$$

Hence, $\{x_k\}$ is a Cauchy sequence and therefore converges to a point $z^* \in \mathcal{Z}$. Since

$$\|x_n - z^*\| \le 2d(x_n),$$

(2.44) implies that the convergence is r-quadratic. □

2.5 Inexact Newton Methods

An *inexact Newton method* [74] uses an approximate Newton step $s = x_+ - x_c$, requiring only that

$$\|\nabla^2 f(x_c)s + \nabla f(x_c)\| \le \eta_c \|\nabla f(x_c)\|, \tag{2.47}$$

i.e., that the linear residual be small. We will refer to any vector s that satisfies (2.47) with $\eta_c < 1$ as an *inexact Newton step*. We will refer to the parameter η_c on the right-hand side of (2.47) as the *forcing term* [99] .

Inexact Newton methods are also called *truncated Newton methods* [75], [198], [199] in the context of optimization. In this book, we consider *Newton–iterative* methods. This is the class of inexact Newton methods in which the linear equation (2.4) for the Newton step is also solved by an iterative method and (2.47) is the termination criterion for that linear iteration. It is standard to refer to the sequence of Newton steps $\{x_n\}$ as the *outer iteration* and the sequence of iterates for the linear equation as the *inner iteration*. The naming convention (see [33], [154], [211]) is that Newton–CG, for example, refers to the Newton–iterative method in which the *conjugate gradient* [141] algorithm is used to perform the inner iteration.

Newton–CG is particularly appropriate for optimization, as we expect positive definite Hessians near a local minimizer. The results for inexact Newton methods from [74] and [154] are sufficient to describe the local convergence behavior of Newton–CG, and we summarize the relevant results from nonlinear equations in §2.5.1. We will discuss the implementation of Newton–CG in §2.5.2.

2.5.1 Convergence Rates

We will prove the simplest of the convergence results for Newton–CG, Theorem 2.5.1, from which Theorem 2.5.2 follows directly. We refer the reader to [74] and [154] for the proof of Theorem 2.5.3.

THEOREM 2.5.1. *Let the standard assumptions hold. Then there are δ and K_I such that if $x_c \in \mathcal{B}(\delta)$, s satisfies* (2.47), *and $x_+ = x_c + s$, then*

$$\|e_+\| \le K_I(\|e_c\| + \eta_c)\|e_c\|. \tag{2.48}$$

Proof. Let δ be small enough so that the conclusions of Lemma 2.3.1 and Theorem 2.3.2 hold. To prove the first assertion (2.48) note that if

$$r = -\nabla^2 f(x_c)s - \nabla f(x_c)$$

is the linear residual, then

$$s + (\nabla^2 f(x_c))^{-1}\nabla f(x_c) = -(\nabla^2 f(x_c))^{-1} r$$

and

$$e_+ = e_c + s = e_c - (\nabla^2 f(x_c))^{-1}\nabla f(x_c) - (\nabla^2 f(x_c))^{-1} r. \tag{2.49}$$

Now, (2.47), (2.7), and (2.6) imply that

$$\begin{aligned} \|s + (\nabla^2 f(x_c))^{-1}\nabla f(x_c)\| &\le \|(\nabla^2 f(x_c))^{-1}\|\eta_c\|\nabla f(x_c)\| \\ &\le 4\kappa(\nabla^2 f(x^*))\eta_c\|e_c\|. \end{aligned}$$

Hence, using (2.49) and Theorem 2.3.2, we have that

$$\begin{aligned} \|e_+\| &\le \|e_c - \nabla^2 f(x_c)^{-1}\nabla f(x_c)\| + 4\kappa(F'(x^*))\eta_c\|e_c\| \\ &\le K\|e_c\|^2 + 4\kappa(\nabla^2 f(x^*))\eta_c\|e_c\|, \end{aligned}$$

where K is the constant from (2.8). If we set

$$K_I = K + 4\kappa(\nabla^2 f(x^*)),$$

the proof is complete. □

THEOREM 2.5.2. *Let the standard assumptions hold. Then there are δ and $\bar{\eta}$ such that if $x_0 \in \mathcal{B}(\delta)$, $\{\eta_n\} \subset [0, \bar{\eta}]$, then the inexact Newton iteration*

$$x_{n+1} = x_n + s_n,$$

where

$$\|\nabla^2 f(x_n) s_n + \nabla f(x_n)\| \le \eta_n \|\nabla f(x_n)\|,$$

converges q-linearly to x^*. Moreover

- *if $\eta_n \to 0$ the convergence is q-superlinear, and*
- *if $\eta_n \le K_\eta \|\nabla f(x_n)\|^p$ for some $K_\eta > 0$ the convergence is q-superlinear with q-order $1 + p$.*

The similarity between Theorem 2.5.2 and Theorem 2.3.5, the convergence result for the chord method, should be clear. Rather than require that the approximate Hessian be accurate, we demand that the linear iteration produce a sufficiently small relative residual. Theorem 2.5.3 is the remarkable statement that any reduction in the relative linear residual will suffice for linear convergence in a certain norm. This statement implies [154] that $\|\nabla f(x_n)\|$ will converge to zero q-linearly, or, equivalently, that $x_n \to x^*$ q-linearly with respect to $\|\cdot\|_*$, which is defined by

$$\|x\|_* = \|\nabla^2 f(x^*) x\|.$$

THEOREM 2.5.3. *Let the standard assumptions hold. Then there is δ such that if $x_c \in \mathcal{B}(\delta)$, s satisfies* (2.47), *$x_+ = x_c + s$, and $\eta_c \le \eta < \bar{\eta} < 1$, then*

$$\|e_+\|_* \le \bar{\eta} \|e_c\|_*. \tag{2.50}$$

THEOREM 2.5.4. *Let the standard assumptions hold. Then there is δ such that if $x_0 \in \mathcal{B}(\delta)$, $\{\eta_n\} \subset [0, \eta]$ with $\eta < \bar{\eta} < 1$, then the inexact Newton iteration*

$$x_{n+1} = x_n + s_n,$$

where

$$\|\nabla^2 f(x_n) s_n + \nabla f(x_n)\| \le \eta_n \|\nabla f(x_n)\|$$

converges q-linearly with respect to $\|\cdot\|_*$ to x^*. Moreover

- *if $\eta_n \to 0$ the convergence is q-superlinear, and*
- *if $\eta_n \le K_\eta \|\nabla f(x_n)\|^p$ for some $K_\eta > 0$ the convergence is q-superlinear with q-order $1 + p$.*

Q-linear convergence of $\{x_n\}$ to a local minimizer with respect to $\|\cdot\|_*$ is equivalent to q-linear convergence of $\{\nabla f(x_n)\}$ to zero. We will use the rate of convergence of $\{\nabla f(x_n)\}$ in our computational examples to compare various methods.

2.5.2 Implementation of Newton–CG

Our implementation of Newton–CG approximately solves the equation for the Newton step with CG. We make the implicit assumption that ∇f has been computed sufficiently accurately for $D_h^2 f(x : w)$ to be a useful approximate Hessian of the Hessian–vector product $\nabla^2 f(x) w$.

Forward Difference CG

Algorithm `fdcg` is an implementation of the solution by CG of the equation for the Newton step (2.4). In this algorithm we take care to identify failure in CG (i.e., detection of a vector p for which $p^T Hp \leq 0$). This failure either means that H is singular ($p^T Hp = 0$; see exercise 2.7.3) or that $p^T Hp < 0$, i.e., p is a *direction of negative curvature*. The algorithms we will discuss in §3.3.7 make good use of directions of negative curvature. The initial iterate for forward difference CG iteration should be the zero vector. In this way the first iterate will give a steepest descent step, a fact that is very useful. The inputs to Algorithm `fdcg` are the current point x, the objective f, the forcing term η, and a limit on the number of iterations $kmax$. The output is the inexact Newton direction d. Note that in step 2b $D_h^2 f(x : p)$ is used as an approximation to $\nabla^2 f(x)p$.

ALGORITHM 2.5.1. $\texttt{fdcg}(x, f, \eta, kmax, d)$

1. $r = -\nabla f(x)$, $\rho_0 = \|r\|_2^2$, $k = 1$, $d = 0$.
2. *Do While* $\sqrt{\rho_{k-1}} > \eta \|\nabla f(x)\|$ *and* $k < kmax$
   (a) *if* $k = 1$ *then* $p = r$
   else
   $\beta = \rho_{k-1}/\rho_{k-2}$ *and* $p = r + \beta p$
   (b) $w = D_h^2 f(x : p)$
   If $p^T w = 0$ *signal indefiniteness; stop.*
   If $p^T w < 0$ *signal negative curvature; stop.*
   (c) $\alpha = \rho_{k-1}/p^T w$
   (d) $d = d + \alpha p$
   (e) $r = r - \alpha w$
   (f) $\rho_k = \|r\|^2$
   (g) $k = k + 1$

Preconditioning can be incorporated into a Newton–CG algorithm by using a forward difference formulation, too. Here, as in [154], we denote the preconditioner by M. Aside from M, the inputs and output of Algorithm `fdpcg` are the same as that for Algorithm `fdcg`.

ALGORITHM 2.5.2. $\texttt{fdpcg}(x, f, M, \eta, kmax, d)$

1. $r = -\nabla f(x)$, $\rho_0 = \|r\|_2^2$, $k = 1$, $d = 0$.
2. *Do While* $\sqrt{\rho_{k-1}} > \eta \|\nabla f(x)\|$ *and* $k < kmax$
   (a) $z = Mr$
   (b) $\tau_{k-1} = z^T r$
   (c) *if* $k = 1$ *then* $\beta = 0$ *and* $p = z$
   else
   $\beta = \tau_{k-1}/\tau_{k-2}$, $p = z + \beta p$
   (d) $w = D_h^2 f(x : p)$
   If $p^T w = 0$ *signal indefiniteness; stop.*
   If $p^T w < 0$ *signal negative curvature; stop.*
   (e) $\alpha = \tau_{k-1}/p^T w$
   (f) $d = d + \alpha p$

(g) $r = r - \alpha w$

(h) $\rho_k = r^T r$

(i) $k = k + 1$

In our formulation of Algorithms `fdcg` and `fdpcg`, indefiniteness is a signal that we are not sufficiently near a minimum for the theory in this section to hold. In §3.3.7 we show how negative curvature can be exploited when far from the solution.

One view of preconditioning is that it is no more than a rescaling of the independent variables. Suppose, rather than (1.2), we seek to solve

$$\min_y \hat{f}(y), \tag{2.51}$$

where $\hat{f}(y) = f(M^{1/2}y)$ and M is spd. If y^* is a local minimizer of $\hat{f}$, then $x^* = M^{1/2}y^*$ is a local minimizer of f and the two problems are equivalent. Moreover, if $x = M^{1/2}y$ and ∇_x and ∇_y denote gradients in the x and y coordinates, then

$$\nabla_y \hat{f}(y) = M^{1/2}\nabla_x f(x)$$

and

$$\nabla_y^2 \hat{f}(y) = M^{1/2}(\nabla_x^2 f(x))M^{1/2}.$$

Hence, the scaling matrix plays the role of the square root of the preconditioner for the preconditioned conjugate gradient algorithm.

Newton–CG

The theory guarantees that if x_0 is near enough to a local minimizer then $\nabla^2 f(x_n)$ will be spd for the entire iteration and x_n will converge rapidly to x^*. Hence, Algorithm `newtcg` will not terminate with failure because of an increase in f or an indefinite Hessian. Note that both the forcing term η and the preconditioner M can change as the iteration progresses.

ALGORITHM 2.5.3. `newtcg`(x, f, τ, η)

1. $r_c = r_0 = \|\nabla f(x)\|$
2. *Do while* $\|\nabla f(x)\| > \tau_r r_0 + \tau_a$

   (a) *Select* η *and a preconditioner* M.

   (b) `fdpcg`$(x, f, M, \eta, kmax, d)$
   If indefiniteness has been detected, terminate with failure.

   (c) $x = x + d$.

   (d) *Evaluate* f *and* $\nabla f(x)$.
   If f *has not decreased, terminate with failure.*

   (e) $r_+ = \|\nabla f(x)\|, \sigma = r_+/r_c, r_c = r_+$.

The implementation of Newton–CG is simple, but, as presented in Algorithm `newtcg`, incomplete. The algorithm requires substantial modification to be able to generate the good initial data that the local theory requires. We return to this issue in §3.3.7.

There is a subtle problem with Algorithm `fdpcg` in that the algorithm is equivalent to the application of the preconditioned conjugate gradient algorithm to the matrix B that is determined by

$$Bp_i = w_i = D_h^2 f(x : p_i), 1 \le i \le N.$$

However, since the map $p \to D_h^2 f(x : p)$ is not linear in p, the quality of B as an approximation to $\nabla^2 f(x)$ may degrade as the linear iteration progresses. Usually this will not cause problems unless many iterations are needed to satisfy the inexact Newton condition. However, if one does not see the expected rate of convergence in a Newton–CG iteration, this could be a factor [128]. One partial remedy is to use a centered-difference Hessian–vector product [162], which reduces the error in B. In exercise 2.7.15 we discuss a more complex and imaginative way to compute accurate Hessians.

2.6 Examples

In this section we used the collection of MATLAB codes but disabled the features (see Chapter 3) that assist in convergence when far from a minimizer. We took care to make certain that the initial iterates were near enough to the minimizer so that the observations of local convergence corresponded to the theory. In practical optimization problems, good initial data is usually not available and the globally convergent methods discussed in Chapter 3 must be used to start the iteration.

The plots in this section have the characteristics of local convergence in that both the gradient norms and function values are decreasing. The reader should contrast this with the examples in Chapter 3.

2.6.1 Parameter Identification

For this example, $M = 100$, and the observations are those of the exact solution with $c = k = 1$, which we computed analytically. We used $T = 10$ and $u_0 = 10$. We computed the displacement and solved the sensitivity equations with the stiff solver `ode15s`. These results could be obtained as well in a FORTRAN environment using, for example, the `LSODE` code [228]. The relative and absolute error tolerances for the integrator were both set to 10^{-8}. In view of the expected accuracy of the gradient, we set the forward difference increment for the approximate Hessian to $h = 10^{-4}$. We terminated the iterations when $\|\nabla f\| < 10^{-4}$. Our reasons for this are that, for the zero residual problem considered here, the standard assumptions imply that $f(x) = O(\|\nabla f(x)\|)$ for x near the solution. Hence, since we can only resolve f to an accuracy of 10^{-8}, iteration beyond the point where $\|\nabla f\| < 10^{-4}$ cannot be expected to lead to a further decrease in f. In fact we observed this in our computations.

The iterations are very sensitive to the initial iterate. We used $x_0 = (1.1, 1.05)^T$; initial iterates much worse than that caused Newton's method to fail. The more robust methods from Chapter 3 should be viewed as essential components of even a simple optimization code.

In Table 2.1 we tabulate the history of the iteration for both the Newton and Gauss–Newton methods. As expected for a small residual problem, Gauss–Newton performs well and, for this example, even converges in fewer iterations. The real benefit of Gauss–Newton is that computation of the Hessian can be avoided, saving considerable computational work by exploiting the structure of the problem. In the computation reported here, the MATLAB `flops` command indicates that the Newton iteration took roughly 1.9 million floating point operations and Gauss–Newton roughly 650 thousand. This difference would be much more dramatic if there were more than two parameters or the cost of an evaluation of f depended on N in a significant way (which it does not in this example).

Figure 2.1 is a graphical representation of the convergence history from Table 2.1. We think that the plots are a more effective way to understand iteration statistics and will present mostly graphs for the remainder of the book. The concavity of the plots of the gradient norms is the signature of superlinear convergence.

Table 2.1: *Parameter identification problem, locally convergent iterations.*

	Newton		Gauss–Newton	
n	$\|\nabla f(x_n)\|$	$f(x_n)$	$\|\nabla f(x_n)\|$	$f(x_n)$
0	2.33e+01	7.88e-01	2.33e+01	7.88e-01
1	6.87e+00	9.90e-02	1.77e+00	6.76e-03
2	4.59e-01	6.58e-04	1.01e-02	4.57e-07
3	2.96e-03	3.06e-08	9.84e-07	2.28e-14
4	2.16e-06	4.15e-14		

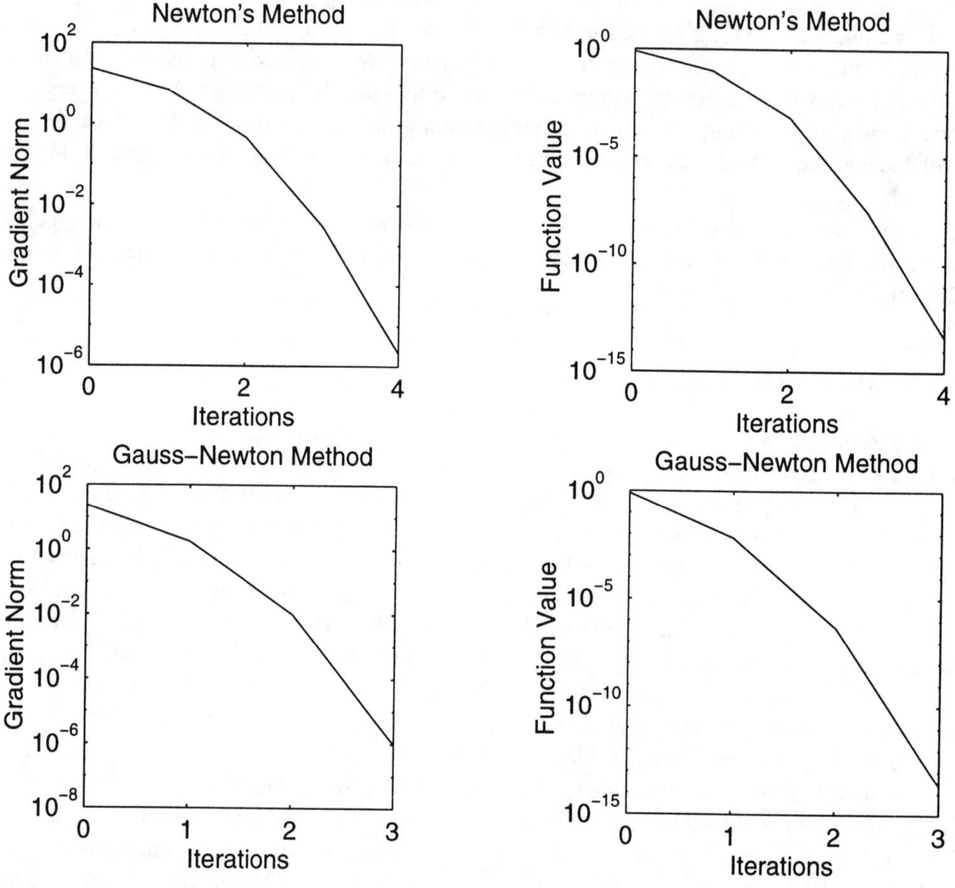

Figure 2.1: *Local Optimization for the Parameter ID Problem*

We next illustrate the difference between Gauss–Newton and Newton on a nonzero residual problem. We use the same example as before with the observations randomly perturbed. We used the MATLAB `rand` function for this, perturbing the samples of the analytic solution by $.5 \times \text{rand}(M, 1)$. The least squares residual is about 3.6 and the plots in Figure 2.2 indicate that Newton's method is still converging quadratically, but the rate of Gauss–Newton is linear. The linear convergence of Gauss–Newton can be seen clearly from the linear semilog plot of the gradient norms. Even so, the Gauss–Newton iteration was more efficient, in terms of floating point operation, than Newton's method. The Gauss–Newton iteration took roughly 1 million floating point operations while the Newton iteration took 1.4 million.

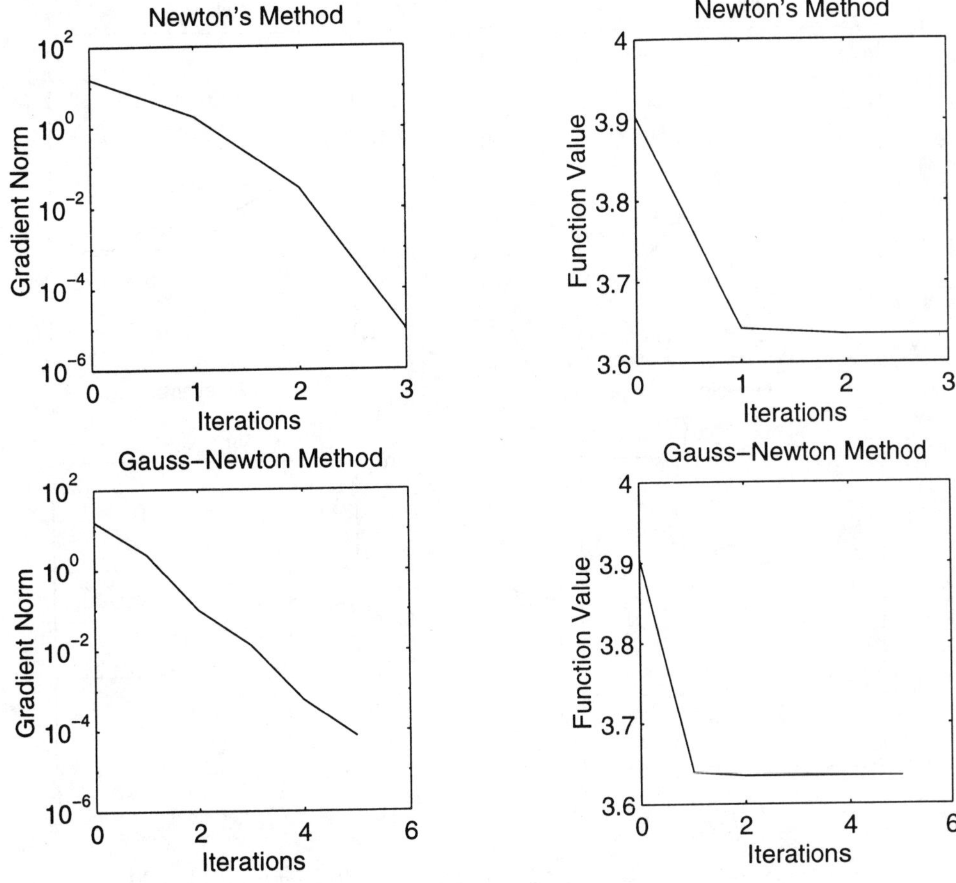

Figure 2.2: *Local Optimization for the Parameter ID Problem, Nonzero Residual*

2.6.2 Discrete Control Problem

We solve the discrete control problem from §1.6.1 with $N = 400$, $T = 1$, $y_0 = 0$,

$$L(y, u, t) = (y - 3)^2 + .5u^2, \text{ and } \phi(y, u, t) = uy + t^2$$

with Newton–CG and two different choices, $\eta = .1, .0001$, of the forcing term. The initial iterate was $u_0 = (10, 10, \dots, 10)^T$ and the iteration was terminated when $\|\nabla f\| < 10^{-8}$. In Figure 2.3 one can see that the small forcing term produces an iteration history with the concavity of superlinear convergence. The limiting q-linear behavior of an iteration with constant η is not yet visible. The iteration with the larger value of η is in the q-linearly convergent stage, as the linear plot of ∇f against the iteration counter shows.

The cost of the computation is not reflected by the number of nonlinear iterations. When $\eta = .0001$, the high accuracy of the linear solve is not rewarded. The computation with $\eta = .0001$ required 8 nonlinear iterations, a total of 32 CG iterations, roughly 1.25 million floating point operations, and 41 gradient evaluations. The optimization with $\eta = .1$ needed 10 nonlinear iterations, a total of 13 CG iterations, roughly 820 thousand floating point operations, and 24 gradient evaluations.

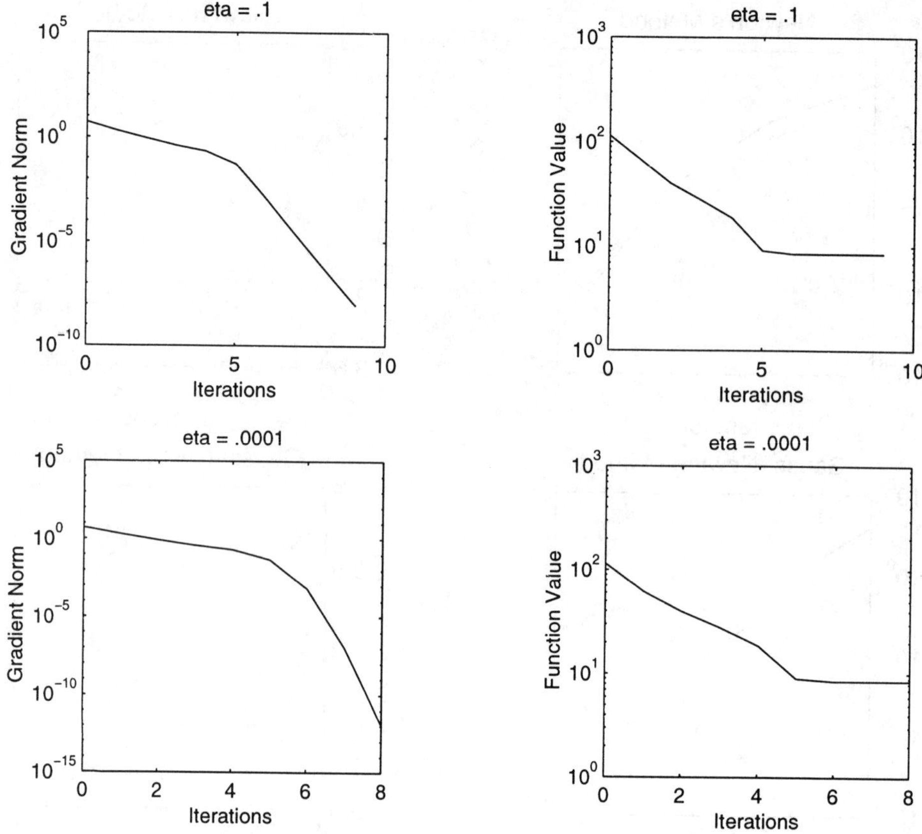

Figure 2.3: *Newton–CG for the Discrete Control Problem:* $\eta = .1, .0001$

2.7 Exercises on Local Convergence

2.7.1. Apply Newton's method with (a) analytic first and second derivatives, (b) analytic first derivatives and forward difference second derivatives, and (c) forward difference first and second derivatives to find a local minimum of

1. $f(x) = \sin^2(x)$,
2. $f(x) = e^{x^2}$, and
3. $f(x) = x^4$.

Use difference steps of $h = 10^{-1}, 10^{-2}, 10^{-4}$, and 10^{-8}. Explain your results.

2.7.2. Repeat part (c) of exercise 2.7.1. Experiment with

$$f(x) = e^{x^2} + 10^{-4}\texttt{rand}(x) \text{ and } f(x) = x^2 + 10^{-4}\texttt{rand}(x),$$

where $rand$ denotes the random number generator in your computing environment. Explain the differences in the results.

2.7.3. Show that if A is symmetric, $p \neq 0$, and $p^T Ap = 0$, then A is either singular or indefinite.

2.7.4. Show that if $b \in R^N$ and the $N \times N$ matrix A is symmetric and has a negative eigenvalue, then the quadratic functional

$$m(x) = x^T Ax + x^T b$$

does not have a minimizer. Show that if u is an eigenvector corresponding to a negative eigenvalue of the Hessian, then u is a direction of negative curvature.

2.7.5. If $N = 1$, the local quadratic model could easily be replaced by a local *quartic* (i.e., fourth-degree) model (what would be wrong with a cubic model?). If a method is based on minimization of the local quartic model, what kind of local convergence would you expect? How would you extend this method to the case $N > 1$? Look at [30] for some results on this.

2.7.6. Show that if the standard assumptions hold, h is sufficiently small, and x is sufficiently near x^*, the difference Hessian defined by (2.19), $\nabla_h^2 f(x)$, is spd.

2.7.7. Write a locally convergent Newton method code based on accurate function and gradient information and forward difference Hessians using (2.19). Be sure that your code tests for positivity of the Hessian so that you can avoid convergence to a local maximum. Is the test for positivity expensive? Apply your code to the parameter ID problem from §1.6.2. If you use an ODE solver that lets you control the accuracy of the integration, try values of the accuracy from 10^{-8} to 10^{-2} and see how the iteration performs. Be sure that your difference Hessian reflects the accuracy in the gradient.

2.7.8. Let the standard assumptions hold and let $\lambda_s > 0$ be the smallest eigenvalue of $\nabla^2 f(x^*)$. Give the best (i.e., largest) bound you can for ρ such that $\nabla^2 f(x)$ is positive definite for all $x \in \mathcal{B}(\rho)$.

2.7.9. Use the definition of $A^\dagger$ to prove (2.41).

2.7.10. Fill in the missing details in the proof of Theorem 2.4.2 by showing how the Kantorovich theorem can be used to prove the existence of $\xi(x)$.

2.7.11. Let $f(x) = x^2$ and $\epsilon_f(x) = \sin(100x)/10$. Using an initial iterate of $x_0 = 1$, try to find a local minimum of $f + \epsilon_f$ using Newton's method with analytic gradients and Hessians. Repeat the experiment with difference gradients and Hessians (try forward differences with a step size of $h = .2$).

2.7.12. Solve the parameter ID problem from §2.6 with the observations perturbed randomly (for example, you could use the MATLAB `rand` function for this). Vary the amplitude of the perturbation and see how the performance of Newton and Gauss–Newton changes.

2.7.13. Derive sensitivity equations for the entries of the Hessian for the parameter ID objective function. In general, if there are P parameters, how many sensitivity equations would need to be solved for the gradient? How many for the Hessian?

2.7.14. Solve the discrete control problem from §2.6.2 using Newton–CG with forcing terms that depend on n. Consider $\eta_n = .5/n$, $\eta_n = \min(.1, \|\nabla f(u_n)\|)$, and some of the choices from [99]. Vary N and the termination criteria and compare the performance with the constant η choice in §2.6.2.

2.7.15. Let $F : R^N \rightarrow R^M$ (where M and N need not be the same) be sufficiently smooth (how smooth is that?) and be such that F can also be computed for complex arguments. Show that [181], [245]

$$\text{Im}(F(x + ihu))/h = F'(x)u + O(h^2),$$

where Im denotes imaginary part. What happens if there is error in F? How can you use this fact to compute better difference gradients and Hessians?

Chapter 3

Global Convergence

The locally convergent algorithms discussed in Chapter 2 can and do fail when the initial iterate is not near the root. The reasons for this failure, as we explain below, are that the Newton direction may fail to be a direction of descent for f and that even when a search direction is a direction of decrease of f, as $-\nabla f$ is, the length of the step can be too long. Hence, taking a Newton (or Gauss–Newton, or inexact Newton) step can lead to an increase in the function and divergence of the iteration (see exercise 3.5.14 for two dramatic examples of this). The *globally convergent* algorithms developed in this chapter partially address this problem by either finding a local minimum or failing in one of a small number of easily detectable ways.

These are not algorithms for global optimization. When these algorithms are applied to problems with many local minima, the results of the iteration may depend in complex ways on the initial iterate.

3.1 The Method of Steepest Descent

The *steepest descent direction* from x is $d = -\nabla f(x)$. The *method of steepest descent* [52] updates the current iteration x_c by the formula

$$x_+ = x_c - \lambda \nabla f(x_c). \tag{3.1}$$

If we take the simple choice $\lambda = 1$, then x_+ is not guaranteed to be nearer a solution than x_c, even if x_c is very near a solution that satisfies the standard assumptions. The reason for this is that, unlike the Newton direction, the steepest descent direction scales with f. The Newton step, on the other hand, is the same for f as it is for cf for any $c \neq 0$ but need not be a direction of decrease for f.

To make the method of steepest descent succeed, it is important to choose the *steplength* λ. One way to do this, which we analyze in §3.2, is to let $\lambda = \beta^m$, where $\beta \in (0,1)$ and $m \geq 0$ is the smallest nonnegative integer such that there is *sufficient decrease* in f. In the context of the steepest descent algorithm, this means that

$$f(x_c - \lambda \nabla f(x_c)) - f(x_c) < -\alpha\lambda \|\nabla f(x_c)\|^2. \tag{3.2}$$

This strategy, introduced in [7] and called the *Armijo rule*, is an example of a *line search* in which one searches on a ray from x_c in a direction in which f is locally decreasing. In (3.2), $\alpha \in (0,1)$ is a parameter, which we discuss after we fully specify the algorithm. This strategy of repeatedly testing for sufficient decrease and reducing the stepsize if the test is failed is called *backtracking* for obvious reasons.

The motivation for (3.2) is that if we approximate f by the *linear model*

$$m_c = f(x_c) + \nabla f(x_c)(x - x_c),$$

then the reduction in the model (i.e., the *predicted reduction* in f) is

$$pred = m_c(x_c) - m_c(x_+) = \lambda \|\nabla f(x_c)\|^2.$$

(3.2) says that the *actual reduction* in f

$$ared = f(x_c) - f(x_+)$$

is at least as much as a fraction of the predicted reduction in the linear model. The parameter α is typically set to 10^{-4}.

The reason we demand sufficient decrease instead of *simple decrease* (i.e., $f(x_+) < f(x_c)$ or $\alpha = 0$) is largely theoretical; a nonzero value of α is required within the proof to insure that the iteration does not stagnate before convergence.

ALGORITHM 3.1.1. $\texttt{steep}(x, f, kmax)$

1. *For* $k = 1, \ldots, kmax$
   (a) *Compute* f *and* ∇f*; test for termination.*
   (b) *Find the least integer* $m \geq 0$ *such that* (3.2) *holds for* $\lambda = \beta^m$.
   (c) $x = x + \lambda d$.
2. *If* $k = kmax$ *and the termination test is failed, signal failure.*

The termination criterion could be based on (2.12), for example.

3.2 Line Search Methods and the Armijo Rule

We introduce a few new concepts so that our proof of convergence of Algorithm steep will also apply to a significantly more general class of algorithms.

DEFINITION 3.2.1. *A vector* $d \in R^N$ *is a descent direction for* f *at* x *if*

$$\left.\frac{df(x+td)}{dt}\right|_{t=0} = \nabla f(x)^T d < 0.$$

Clearly the steepest descent direction $d = -\nabla f(x)$ is a descent direction. A *line search algorithm* searches for decrease in f in a descent direction, using the Armijo rule for stepsize control, unless $\nabla f(x) = 0$.

We will consider descent directions based on *quadratic models* of f of the form

$$m(x) = f(x_c) + \nabla f(x_c)^T(x - x_c) + \frac{1}{2}(x - x_c)^T H_c(x - x_c),$$

where H_c, which is sometimes called the *model Hessian,* is spd. We let $d = x - x_c$ be such that $m(x)$ is minimized. Hence,

$$\nabla m(x) = \nabla f(x_c) + H_c(x - x_c) = 0$$

and hence

$$d = -H_c^{-1}\nabla f(x_c). \tag{3.3}$$

The steepest descent direction satisfies (3.3) with $H_c = I$. However, the Newton direction $d = -\nabla^2 f(x)^{-1}\nabla f(x)$ may fail to be a descent direction if x is far from a minimizer because

$\nabla^2 f$ may not be spd. Hence, unlike the case for nonlinear equations [154], Newton's method is not a generally good global method, even with a line search, and must be modified (see [113], [117], [231], and [100]) to make sure that the model Hessians are spd.

The algorithm we analyze in this section is an extension of Algorithm `steep` that allows for descent directions that satisfy (3.3) for spd H. We modify (3.2) to account for H and the new descent direction $d = -H^{-1}\nabla f(x)$. The general *sufficient decrease* condition is

$$f(x_c + \lambda d) - f(x_c) < \alpha\lambda\nabla f(x_c)^T d. \tag{3.4}$$

Here, as in (3.2), $\alpha \in (0, 1)$ is an algorithmic parameter. Typically $\alpha = 10^{-4}$.

The stepsize reduction scheme in step 1b of Algorithm `steep` is crude. If β is too large, too many stepsize reductions may be needed before a step is accepted. If β is too small, the progress of the entire iteration may be retarded. We will address this problem in two ways. In §3.2.1 we will construct polynomial models of f along the descent direction to predict an optimal factor by which to reduce the step. In §3.3.3 we describe a method which remembers the steplength from the previous iteration.

Our proofs require only the following general line search strategy. If a steplength λ_c has been rejected (i.e., (3.4) fails with $\lambda = \lambda_c$), construct

$$\lambda_+ \in [\beta_{low}\lambda_c, \beta_{high}\lambda_c], \tag{3.5}$$

where $0 < \beta_{low} \leq \beta_{high} < 1$. The choice $\beta = \beta_{low} = \beta_{high}$ is the simple rule in Algorithm `steep`. An *exact line search*, in which λ is the exact minimum of $f(x_c + \lambda d)$, is not only not worth the extra expense but can degrade the performance of the algorithm.

ALGORITHM 3.2.1. `optarm`$(x, f, kmax)$

1. *For* $k = 1, \ldots, kmax$
   (a) *Compute* f *and* ∇f; *test for termination.*
   (b) *Construct an spd matrix* H *and solve* (3.3) *to obtain a descent direction* d.
   (c) *Beginning with* $\lambda = 1$, *repeatedly reduce* λ *using any strategy that satisfies* (3.5) *until* (3.4) *holds.*
   (d) $x = x + \lambda d$.
2. *If* $k = kmax$ *and the termination test is failed, signal failure.*

In the remainder of this section we prove that if the sequence of model Hessians remains uniformly bounded and positive definite and the sequence of function values $\{f(x_k)\}$ is bounded from below, then any limit point of the sequence $\{x_k\}$ generated by Algorithm `optarm` converges to a point x^* that satisfies the necessary conditions for optimality. We follow that analysis with a local convergence theory that is much less impressive than that for Newton's method.

We begin our analysis with a simple estimate that follows directly from the spectral theorem for spd matrices.

LEMMA 3.2.1. *Let* H *be spd with smallest and largest eigenvalues* $0 < \lambda_s < \lambda_l$. *Then for all* $z \in R^N$,

$$\lambda_l^{-1}\|z\|^2 \leq z^T H^{-1} z \leq \lambda_s^{-1}\|z\|^2.$$

The first step in the analysis is to use Lemma 3.2.1 to obtain a lower bound for the steplength.

LEMMA 3.2.2. *Assume that* ∇f *is Lipschitz continuous with Lipschitz constant* L. *Let* $\alpha \in (0, 1)$, $x \in R^N$, *and* H *be an spd matrix. Let* $\lambda_s > 0$ *be the smallest and* $\lambda_l \geq \lambda_s$ *the* *largest eigenvalues of H. Let d be given by* (3.3). *Assume that $\nabla f(x) \neq 0$. Then* (3.4) *holds for any λ such that*

$$0 < \lambda \leq \frac{2\lambda_s(1-\alpha)}{L\kappa(H)}. \tag{3.6}$$

Proof. Let $d = -H^{-1}\nabla f(x)$. By the fundamental theorem of calculus

$$f(x+\lambda d) - f(x) = \int_0^1 \nabla f(x+t\lambda d)^T \lambda d\, dt.$$

Hence

$$\begin{aligned} f(x+\lambda d) &= f(x) + \lambda \nabla f(x)^T d \\ &\quad + \lambda \textstyle\int_0^1 (\nabla f(x+t\lambda d) - \nabla f(x))^T d\, dt. \end{aligned} \tag{3.7}$$

Therefore,

$$f(x+\lambda d) = f(x - \lambda H^{-1}\nabla f(x)) \leq f(x) + \lambda \nabla f(x)^T d + \frac{\lambda^2 L}{2}\|d\|^2.$$

Positivity of H, Lemma 3.2.1, and $\kappa(H) = \lambda_l \lambda_s^{-1}$ imply that

$$\begin{aligned} \|d\|^2 &= \|H^{-1}\nabla f(x)\|^2 \leq \lambda_s^{-2} \nabla f(x)^T \nabla f(x) \\ &\leq -\lambda_l \lambda_s^{-2} \nabla f(x)^T d = -\kappa(H)\lambda_s^{-1}\nabla f(x)^T d. \end{aligned}$$

Hence

$$f(x+\lambda d) \leq f(x) + (\lambda - \lambda^2 L \lambda_s^{-1}\kappa(H)/2)\nabla f(x)^T d,$$

which implies (3.4) if

$$\alpha \leq (1 - \lambda L \lambda_s^{-1}\kappa(H)/2).$$

This is equivalent to (3.6). □

LEMMA 3.2.3. *Let ∇f be Lipschitz continuous with Lipschitz constant L. Let $\{x_k\}$ be the iteration given by Algorithm* `optarm` *with spd matrices H_k that satisfy*

$$\kappa(H_k) \leq \bar{\kappa} \tag{3.8}$$

for all k. Then the steps

$$s_k = x_{k+1} - x_k = \lambda_k d_k = -\lambda_k H_k^{-1}\nabla f(x_k)$$

satisfy

$$\lambda_k \geq \bar{\lambda} = \frac{2\beta_{low}\lambda_s(1-\alpha)}{L\bar{\kappa}} \tag{3.9}$$

and at most

$$m = \log\left(\frac{2\lambda_s(1-\alpha)}{L\bar{\kappa}}\right) / \log(\beta_{high}) \tag{3.10}$$

stepsize reductions will be required.

Proof. In the context of Algorithm `optarm`, Lemma 3.2.2 implies that the line search will terminate when

$$\lambda \leq \frac{2\lambda_s(1-\alpha)}{L\kappa(H_k)},$$

if not before. The most that one can overshoot this is by a factor of β_{low}, which proves (3.9). The line search will require at most m stepsize reductions, where m is the least nonnegative integer such that

$$\frac{2\lambda_s(1-\alpha)}{L\kappa(H_k)} > \beta_{high}^m.$$

This implies (3.10). □

The convergence theorem for Algorithm `optarm` says that if the condition numbers of the matrices H and the norms of the iterates remain bounded, then every limit point of the iteration is a stationary point. Boundedness of the sequence of iterates implies that there will be limit points, but there is no guarantee that there is a unique limit point.

THEOREM 3.2.4. *Let ∇f be Lipschitz continuous with Lipschitz constant L. Assume that the matrices H_k are spd and that there are $\bar{\kappa}$ and λ_l such that $\kappa(H_k) \le \bar{\kappa}$, and $\|H_k\| \le \lambda_l$ for all k. Then either $f(x_k)$ is unbounded from below or*

$$\lim_{k\to\infty} \nabla f(x_k) = 0 \tag{3.11}$$

and hence any limit point of the sequence of iterates produced by Algorithm `optarm` *is a stationary point.*

In particular, if $f(x_k)$ is bounded from below and $x_{k_l} \to x^*$ is any convergent subsequence of $\{x_k\}$, then $\nabla f(x^*) = 0$.

Proof. By construction, $f(x_k)$ is a decreasing sequence. Therefore, if $f(x_k)$ is bounded from below, then $\lim_{k\to\infty} f(x_k) = f^*$ exists and

$$\lim_{k\to\infty} f(x_{k+1}) - f(x_k) = 0. \tag{3.12}$$

By (3.4) and Lemma 3.2.3 we have

$$\begin{aligned} f(x_{k+1}) - f(x_k) &< -\alpha\lambda_k \nabla f(x_k)^T H_k^{-1} \nabla f(x_k) \\ &\le -\alpha\bar{\lambda}\lambda_l^{-1}\|\nabla f(x_k)\|^2 \le 0. \end{aligned}$$

Hence, by (3.12)

$$\|\nabla f(x_k)\|^2 \le \frac{\lambda_l(f(x_k) - f(x_{k+1}))}{\alpha\bar{\lambda}} \to 0$$

as $k \to \infty$. This completes the proof. □

The analysis of the Armijo rule is valid for other line search methods [84], [125], [272], [273]. The key points are that the sufficient decrease condition can be satisfied in finitely many steps and that the stepsizes are bounded away from zero.

3.2.1 Stepsize Control with Polynomial Models

Having computed a descent direction d from x_c, one must decide on a stepsize reduction scheme for iterations in which (3.4) fails for $\lambda = 1$. A common approach [73], [84], [114], [197], [117] is to model

$$\xi(\lambda) = f(x_c + \lambda d)$$

by a cubic polynomial. The data on hand initially are

$$\xi(0) = f(x_c), \xi'(0) = \nabla f(x_c)^T d < 0, \text{ and } \xi(1) = f(x+d),$$

which is enough to form a quadratic model of ξ. So, if (3.4) does not hold with $\lambda = \lambda_0 = 1$, i.e.,

$$\xi(1) = f(x_c + d) \ge f(x_c) + \alpha\nabla f(x_c)^T d = \xi(0) + \alpha\xi'(0),$$

we approximate ξ by the quadratic polynomial

$$q(\lambda) = \xi(0) + \xi'(0)\lambda + (\xi(1) - \xi(0) - \xi'(0))\lambda^2$$

and let λ_1 be the minimum of q on the interval $[\beta_{low}, \beta_{high}] \subset (0,1)$. This minimum can be computed directly since $\alpha \in (0,1)$ and failure of (3.4) imply

$$q''(\lambda) = 2(\xi(1) - \xi(0) - \xi'(0)) > 2(\alpha - 1)\xi'(0) > 0.$$

Therefore, the global minimum of q is

$$\lambda_t = \frac{-\xi'(0)}{2(\xi(1) - \xi(0) - \xi'(0))}.$$

So

$$\lambda_+ = \begin{cases} \beta_{low}, & \lambda_t \le \beta_{low}, \\ \lambda_t, & \beta_{low} < \lambda_t < \beta_{high}, \\ \beta_{high}, & \lambda_t \ge \beta_{high}. \end{cases} \tag{3.13}$$

If our first reduced value of λ does not satisfy (3.4), we base additional reductions on the data

$$\xi(0) = f(x_c), \xi'(0) = \nabla f(x_c)^T d, \xi(\lambda_-), \xi(\lambda_c),$$

where $\lambda_c < \lambda_-$ are the most recent values of λ to fail to satisfy (3.4). This is sufficient data to approximate ξ with a cubic polynomial

$$q(\lambda) = \xi(0) + \xi'(0)\lambda + c_2\lambda^2 + c_3\lambda^3,$$

where c_2 and c_3 can be determined by the equations

$$\begin{aligned} q(\lambda_c) &= \xi(\lambda_c) = f(x_c + \lambda_c d), \\ q(\lambda_-) &= \xi(\lambda_-) = f(x_c + \lambda_- d), \end{aligned}$$

which form the nonsingular linear system for c_2 and c_3

$$\begin{pmatrix} \lambda_c^2 & \lambda_c^3 \\ \lambda_-^2 & \lambda_-^3 \end{pmatrix} \begin{pmatrix} c_2 \\ c_3 \end{pmatrix} = \begin{pmatrix} \xi(\lambda_c) - \xi(0) - \xi'(0)\lambda_c \\ \xi(\lambda_-) - \xi(0) - \xi'(0)\lambda_- \end{pmatrix}. \tag{3.14}$$

As with the quadratic case, q has a local minimum [84] at

$$\lambda_t = \frac{-c_2 + \sqrt{c_2^2 - 3c_3\xi'(0)}}{3c_3}. \tag{3.15}$$

With λ_t in hand, we compute λ_+ using (3.13). The application of (3.13) is called *safeguarding* and is important for the theory, as one can see from the proof of Theorem 3.2.4. Safeguarding is also important in practice because, if the cubic model is poor, the unsafeguarded model can make steplength reductions that are so small that the iteration can stagnate or so large (i.e., near 1) that too many reductions are needed before (3.4) holds.

3.2.2 Slow Convergence of Steepest Descent

Unfortunately, methods based on steepest descent do not enjoy good local convergence properties, even for very simple functions. To illustrate this point we consider the special case of *convex quadratic* objective functions

$$f(x) = \frac{1}{2} x^T A x - b^T x + a,$$

where A is spd, $b \in R^N$, and a is a scalar. We will look at a very simple example, using the method of steepest descent with $H_k = I$ (so $\lambda_l = \lambda_s = 1$) and show how the speed of convergence depends on conditioning and scaling.

LEMMA 3.2.5. *Let f be a convex quadratic and let $H_k = I$ for all k. Then the sequence $\{x_k\}$ generated by Algorithm* `optarm` *converges to the unique minimizer of f.*

Proof. In exercise 3.5.4 you are asked to show that f is bounded from below and that $\nabla f(x) = Ax - b$. Hence $\nabla f(x^*)$ vanishes only at $x^* = A^{-1}b$. Since $\nabla^2 f(x) = A$ is spd, the second-order sufficient conditions hold and x^* is the unique minimizer of f.

Theorem 3.2.4 implies that

$$\lim_{k \to \infty} \nabla f(x_k) = Ax_k - b = A(x_k - x^*) = 0,$$

and hence $x_k \to x^*$. $\square$

Since the steepest descent iteration converges to the unique minimizer of a convex quadratic, we can investigate the rate of convergence without concern about the initial iterate. We do this in terms of the A-norm. The problems can be illustrated with the simplest case $a = 0$ and $b = 0$.

PROPOSITION 3.2.6. *Let $f(x) = x^T Ax/2$ and let $\{x_k\}$ be given by Algorithm* `optarm` *with $H_k = I$ for all k. Then the sequence $\{x_k\}$ satisfies*

$$\|x_{k+1}\|_A = (1 - O(\kappa(A)^{-2}))\|x_k\|_A. \tag{3.16}$$

Proof. The sufficient decrease condition, (3.4), implies that for all k

$$\begin{aligned} x_{k+1}^T A x_{k+1} - x_k^T A x_k &= 2(f(x_{k+1}) - f(x_k)) \\ &\leq 2\alpha \nabla f(x_k)^T (x_{k+1} - x_k) \qquad (3.17) \\ &= 2\alpha\lambda_k \nabla f(x_k)^T d = -2\alpha\lambda_k (Ax_k)^T (Ax_k). \end{aligned}$$

The Lipschitz constant of ∇f is simply $\lambda_l = \|A\|$; hence we may write (3.9) as

$$\lambda_k \geq \bar{\lambda} = \frac{2\beta(1-\alpha)}{\lambda_l \kappa(A)}. \tag{3.18}$$

In terms of the A-norm, (3.17) can be written as

$$\|x_{k+1}\|_A^2 - \|x_k\|_A^2 \leq -2\alpha\bar{\lambda}\lambda_s \|x_k\|_A^2,$$

where we use the fact that

$$\|Az\|^2 = (Az)^T (Az) \geq \lambda_s z^T Az = \lambda_s \|z\|_A^2.$$

Hence,

$$\|x_{k+1}\|_A^2 \le (1 - 2\alpha\bar{\lambda}\lambda_s)\|x_k\|_A^2 \le (1 - 4\alpha(1-\alpha)\beta\kappa(A)^{-2})\|x_k\|_A^2.$$

This completes the proof. □

Now we consider two specific examples. Let $N = 1$ and define

$$f(x) = \omega x^2/2,$$

where

$$\omega < 2(1-\alpha). \tag{3.19}$$

In this case $x^* = 0$. We have $\nabla f(x) = f'(x) = \omega x$ and hence for all $x \in R$,

$$\begin{aligned} f(x - \nabla f(x)) - f(x) &= \frac{\omega x^2}{2}((1-\omega)^2 - 1) \\ &= \frac{\omega^2 x^2}{2}(\omega - 2) \\ &< -\alpha|f'(x)|^2 = -\alpha\omega^2 x^2 \end{aligned}$$

because (3.19) implies that

$$\omega - 2 < -2\alpha.$$

Hence (3.4) holds with $d = \nabla f(x)$ and $\lambda = 1$ for all $x \in R$. The rate of convergence can be computed directly since

$$x_+ = (1-\omega)x_c$$

for all x_c. The convergence is q-linear with q-factor $1-\omega$. So if ω is very small, the convergence will be extremely slow.

Similarly, if ω is large, we see that

$$f(x - \lambda\nabla f(x)) - f(x) = \frac{\omega^2 x^2}{2}(\lambda\omega - 2) < -\alpha\lambda\omega^2 x^2$$

only if

$$\lambda < \frac{2(1-\alpha)}{\omega}.$$

So

$$\beta\frac{2(1-\alpha)}{\omega} < \beta^m = \lambda < \frac{2(1-\alpha)}{\omega}.$$

If ω is very large, many steplength reductions will be required with each iteration and the line search will be very inefficient.

These are examples of *poor scaling*, where a change in f by a multiplicative factor can dramatically improve the efficiency of the line search or the convergence speed. In fact, if $\omega = 1$, steepest descent and Newton's method are the same and only one iteration is required.

The case for a general convex quadratic is similar. Let λ_l and λ_s be the largest and smallest eigenvalues of the spd matrix A. We assume that $b = 0$ and $a = 0$ for this example. We let u_l and u_s be unit eigenvectors corresponding to the eigenvalues λ_l and λ_s. If

$$\lambda_s < 2(1-\alpha)$$

is small and the initial iterate is in the direction of u_s, convergence will require a very large number of iterations (slow). If λ_l is large and the initial iterate is in the direction of u_l, the line search will be inefficient (many stepsize reductions at each iteration).

Newton's method does not suffer from poor scaling of f and converges rapidly with no need for a line search when the initial iterate is near the solution. However, when far away from the solution, the Newton direction may not be a descent direction at all and the line search may fail. Making the transition from steepest descent, which is a good algorithm when far from the solution, to Newton's or some other superlinearly convergent method as the iteration moves toward the solution, is the central problem in the design of line search algorithms. The scaling problems discussed above must also be addressed, even when far from the solution.

3.2.3 Damped Gauss–Newton Iteration

As we showed in §2.4, the steepest descent direction for the overdetermined least squares objective

$$f(x) = \frac{1}{2}\sum_{i=1}^{M} \|r_i(x)\|_2^2 = \frac{1}{2}R(x)^T R(x)$$

is

$$-\nabla f(x) = -R'(x)^T R(x).$$

The steepest descent algorithm could be applied to nonlinear least squares problems with the good global performance and poor local convergence that we expect.

The Gauss–Newton direction at x

$$d^{GS} = -(R'(x)^T R'(x))^{-1} R'(x)^T R(x)$$

is not defined if R' fails to have full column rank. If R' does have full column rank, then

$$(d^{GS})^T \nabla f(x) = -(R'(x)^T R(x))^T (R'(x)^T R'(x))^{-1} R'(x)^T R(x) < 0,$$

and the Gauss–Newton direction is a descent direction. The combination of the Armijo rule with the Gauss–Newton direction is called *damped Gauss–Newton* iteration.

A problem with the damped Gauss–Newton algorithm is that, in order for Theorem 3.2.4 to be applicable, the matrices $\{R'(x_k)^T R'(x_k)\}$ must not only have full column rank but also must be uniformly bounded and well conditioned, which are very strong assumptions (but if they are satisfied, damped Gauss–Newton is a very effective algorithm).

The *Levenberg–Marquardt* method [172], [183] addresses these issues by adding a regularization parameter $\nu > 0$ to $R'(x_c)^T R'(x_c)$ to obtain $x_+ = x_c + s$ where

$$s = -(\nu_c I + R'(x_c)^T R'(x_c))^{-1} R'(x_c)^T R(x_c), \tag{3.20}$$

where I is the $N \times N$ identity matrix. The matrix $\nu_c I + R'(x_c)^T R'(x_c)$ is positive definite. The parameter ν is called the *Levenberg–Marquardt parameter*.

It is not necessary to compute $R'(x_c)^T R'(x_c)$ to compute a Levenberg–Marquardt step [76]. One can also solve the full-rank $(M + N) \times N$ linear least squares problem

$$\min \frac{1}{2}\left\| \begin{bmatrix} R'(x_c) \\ \sqrt{\nu_c} I \end{bmatrix} s + \begin{bmatrix} R(x_c) \\ 0 \end{bmatrix} \right\|^2 \tag{3.21}$$

to compute s (see exercise 3.5.6). Compare this with computing the undamped Gauss–Newton step by solving (2.33).

If one couples the Levenberg–Marquardt method with the Armijo rule, then Theorem 3.2.4 is applicable far from a minimizer and Theorem 2.4.1 nearby. We ask the reader to provide the details of the proof in exercise 3.5.7.

THEOREM 3.2.7. *Let R' be Lipschitz continuous. Let x_k be the Levenberg–Marquardt–Armijo iterates. Assume that $\|R'(x_k)\|$ is uniformly bounded and that the sequence of Levenberg–Marquardt parameters $\{\nu_k\}$ is such that*

$$\kappa(\nu_k I + R'(x_k)^T R'(x_k))$$

is bounded. Then

$$\lim_{k\to\infty} R'(x_k)^T R(x_k) = 0.$$

Moreover, if x^* is any limit point of $\{x_k\}$ at which $R(x^*) = 0$, Assumption 2.4.1 *holds, and $\nu_k \to 0$, then $x_k \to x^*$ q-superlinearly. If, moreover,*

$$\nu_k = O(\|R(x_k)\|)$$

as $k \to \infty$ then the convergence is q-quadratic.

For example, if $\kappa(R'(x_k)^T R'(x_k))$ and $\|R'(x_k)\|$ are bounded then $\nu_k = \min(1, \|R(x_k)\|)$ would satisfy the assumptions of Theorem 3.2.7. For a zero residual problem, this addresses the potential conditioning problems of the damped Gauss–Newton method and still gives quadratic convergence in the terminal phase of the iteration. The Levenberg–Marquardt–Armijo iteration will also converge, albeit slowly, for a large residual problem.

We will not discuss globally convergent methods for underdetermined least squares problems in this book. We refer the reader to [24], [252], and [253] for discussion of underdetermined problems.

3.2.4 Nonlinear Conjugate Gradient Methods

Operationally, the conjugate gradient iteration for a quadratic problem updates the current iteration with a linear combination of the current residual r and a search direction p. The search direction is itself a linear combination of previous residuals. Only r and p need be stored to continue the iteration. The methods discussed in this section seek to continue this idea to more nonlinear problems.

Nonlinear conjugate gradient algorithms have the significant advantage of low storage over most of the other algorithms covered in this book, the method of steepest descent being the exception. For problems so large that the Newton or quasi-Newton methods cannot be implemented using the available storage, these methods are among the few options (see [177] and [5] for examples).

Linear conjugate gradient seeks to minimize $f(x) = x^T Hx/2 - x^T b$. The residual $r = b - Hx$ is simply $-\nabla f(x)$, leading to a natural extension to nonlinear problems in which $r_0 = p_0 = \nabla f(x_0)$ and, for $k \geq 1$,

$$r_k = \nabla f(x_k) \text{ and } p_k = r_k + \beta_k p_{k-1}. \tag{3.22}$$

The update of x

$$x_{k+1} = x_k + \alpha_k p_k$$

can be done with a simple analytic minimization in the quadratic case, but a line search will be needed in the nonlinear case. The missing pieces, therefore, are the choice of β_k, the way the line search will be done, and convergence theory. Theory is needed, for example, to decide if p_k is a descent direction for all k.

The general form of the algorithm is very simple. The inputs are an initial iterate, which will be overwritten by the solution, the function to be minimized, and a termination vector $\tau = (\tau_r, \tau_a)$ of relative and absolute residuals.

ALGORITHM 3.2.2. $\texttt{nlcg}(x, f, \tau)$

1. $r_0 = \|\nabla f(x)\|$; $k = 0$
2. *Do while* $\|\nabla f(x)\| > \tau_r r_0 + \tau_a$
   (a) *If* $k = 0$ *then* $p = -\nabla f(x)$ *else*
   $p = -\nabla f(x) + \beta p$
   (b) $x = x + \alpha p$

The two most common choices for β, both of which are equivalent to the linear CG formula in the quadratic case, are the Fletcher–Reeves [106]

$$\beta_k^{FR} = \frac{\|\nabla f(x_k)\|^2}{\|\nabla f(x_{k-1})\|^2} \tag{3.23}$$

and Polak–Ribière [215], [216]

$$\beta_k = \frac{\nabla f(x_k)^T(\nabla f(x_k) - \nabla f(x_{k-1}))}{\|\nabla f(x_{k-1})\|^2} \tag{3.24}$$

formulas. The Fletcher–Reeves method has been observed to take long sequences of very small steps and virtually stagnate [112], [207], [208], [226]. The Polak–Ribière formula performs much better and is more commonly used but has a less satisfactory convergence theory.

The line search has more stringent requirements, at least for proofs of convergence, than are satisfied by the Armijo method that we advocate for steepest descent. We require that the steplength parameter satisfies the *Wolfe conditions* [272], [273]

$$f(x_k + \alpha_k p_k) \leq f(x_k) + \sigma_\alpha \alpha_k \nabla f(x_k)^T p_k \tag{3.25}$$

and

$$\nabla f(x_k + \alpha_k p_k)^T p_k \geq \sigma_\beta \nabla f(x_k)^T p_k, \tag{3.26}$$

where $0 < \sigma_\alpha < \sigma_\beta < 1$. The first of the Wolfe conditions (3.25) is the sufficient decrease condition, (3.4), that all line search algorithms must satisfy. The second (3.26) is often, but not always, implied by the Armijo backtracking scheme of alternating a test for sufficient decrease and reduction of the steplength. One can design a line search method that will, under modest assumptions, find a steplength satisfying the Wolfe conditions [104], [171], [193].

The convergence result [3] for the Fletcher–Reeves formula requires a bit more. The proof that p_k is descent direction requires the *strong Wolfe conditions*, which replace (3.26) by

$$|\nabla f(x_k + \alpha_k p_k)^T p_k| \leq -\sigma_\beta \nabla f(x_k)^T p_k \tag{3.27}$$

and demand that $0 < \sigma_\alpha < \sigma_\beta < 1/2$. The algorithm from [193], for example, will find a point satisfying the strong Wolfe conditions.

THEOREM 3.2.8. *Assume that the set*

$$\mathcal{N} = \{x \,|\, f(x) \leq f(x_0)\}$$

is bounded and that f is Lipschitz continuously differentiable in a neighborhood of $\mathcal{N}$. Let Algorithm `nlcg` *be implemented with the Fletcher–Reeves formula and a line search that satisfies the strong Wolfe conditions. Then*

$$\lim \nabla f(x_k) = 0.$$

This result has been generalized to allow for any choice of β_k such that $|\beta_k| \le \beta_k^{FR}$ [112]. A similar result for the Polak–Ribière method, but with more complex conditions on the line search, has been proved in [134]. This complexity in the line search is probably necessary, as there are examples where reasonable line searches lead to failure in the Polak–Ribière method, [222]. One can also prove convergence if β_k^{PR} is replaced by $\max(\beta_k^{PR}, 0)$ [112].

There is continuing research on these methods and we point to [112], [134], [205], and [202] as good sources.

3.3 Trust Region Methods

Trust region methods overcome the problems that line search methods encounter with non-spd approximate Hessians. In particular, a Newton trust region strategy allows the use of complete Hessian information even in regions where the Hessian has negative curvature. The specific trust region methods we will present effect a smooth transition from the steepest descent direction to the Newton direction in a way that gives the global convergence properties of steepest descent and the fast local convergence of Newton's method.

The idea is very simple. We let Δ be the radius of the ball about x_c in which the quadratic model

$$m_c(x) = f(x_c) + \nabla f(x_c)^T(x - x_c) + (x - x_c)^T H_c(x - x_c)/2$$

can be trusted to accurately represent the function. Δ is called the *trust region radius* and the ball

$$\mathcal{T}(\Delta) = \{x \mid \|x - x_c\| \le \Delta\}$$

is called the *trust region*.

We compute the new point x_+ by (approximately) minimizing m_c over $\mathcal{T}(\Delta)$. The *trust region problem* for doing that is usually posed in terms of the difference s_t between x_c and the minimizer of m_c in the trust region

$$\min_{\|s\| \le \Delta} m_c(x_c + s). \tag{3.28}$$

We will refer to either the trial step s_t or the trial solution $x_t = x_c + s_t$ as the solution to the trust region problem.

Having solved the trust region problem, one must decide whether to accept the step and/or to change the trust region radius. The trust region methods that we will discuss in detail approximate the solution of the trust region problem with the minimizer of the quadratic model along a piecewise linear path contained in the trust region. Before discussing these specific methods, we present a special case of a result from [223] on global convergence.

A prototype trust region algorithm, upon which we base the specific instances that follow, is Algorithm 3.3.1.

ALGORITHM 3.3.1. `trbasic`(x, f)

1. *Initialize the trust region radius* Δ.
2. *Do until termination criteria are satisfied*
   (a) *Approximately solve the trust region problem to obtain* x_t.
   (b) *Test both the trial point and the trust region radius and decide whether or not to accept the step, the trust region radius, or both. At least one of* x *or* Δ *will change in this phase.*

Most trust region algorithms differ only in how step 2a in Algorithm `trbasic` is done. There are also different ways to implement step 2b, but these differ only in minor details and the approach we present next in §3.3.1 is very representative.

3.3.1 Changing the Trust Region and the Step

The trust region radius and the new point are usually tested simultaneously. While a notion of sufficient decrease is important, the test is centered on how well the quadratic model approximates the function inside the trust region. We measure this by comparing the *actual reduction* in f

$$ared = f(x_c) - f(x_t)$$

with the *predicted reduction*, i.e., the decrease in the quadratic model

$$pred = m_c(x_c) - m_c(x_t) = -\nabla f(x_c)^T s_t - s_t^T H_c s_t/2.$$

$pred > 0$ for all the trust region algorithms we discuss in this book unless $\nabla f(x_c) = 0$. We will introduce three control parameters

$$\mu_0 \le \mu_{low} < \mu_{high},$$

which are used to determine if the trial step should be rejected ($ared/pred < \mu_0$) and/or the trust region radius should be decreased ($ared/pred < \mu_{low}$), increased ($ared/pred > \mu_{high}$), or left unchanged. Typical values are .25 for μ_{low} and .75 for μ_{high}. Both $\mu_0 = 10^{-4}$ or $\mu_0 = \mu_{low}$ are used. One can also use the sufficient decrease condition (3.4) to determine if the trial step should be accepted [84].

We will contract and expand the trust region radius by simply multiplying Δ by constants

$$0 < \omega_{down} < 1 < \omega_{up}.$$

Typical values are $\omega_{down} = 1/2$ and $\omega_{up} = 2$. There are many other ways to implement a trust region adjustment algorithm that also give global convergence. For example, the relative error $|pred - ared|/\|\nabla f\|$ can be used [84] rather than the ratio $ared/pred$. Finally we limit the number of times the trust region radius can be expanded by requiring

$$\Delta \le C_T \|\nabla f(x_c)\|, \tag{3.29}$$

for some $C_T > 1$, which may depend on x_c. This only serves to eliminate the possibility of infinite expansion and is used in the proofs. Many of the dogleg methods which we consider later automatically impose (3.29).

The possibility of expansion is important for efficiency in the case of poor scaling of f. The convergence theory presented here [162] also uses the expansion phase in the proof of convergence, but that is not essential. We will present the algorithm to test the trust region in a manner, somewhat different from much of the literature, that only returns once a new iterate has been accepted.

ALGORITHM 3.3.2. $\texttt{trtest}(x_c, x_t, x_+, f, \Delta)$

1. $z = x_c$

2. *Do while* $z = x_c$

   (a) $ared = f(x_c) - f(x_t)$, $s_t = x_t - x_c$, $pred = -\nabla f(x_c)^T s_t - s_t^T H_c s_t/2$

   (b) *If* $ared/pred < \mu_0$ *then set* $z = x_c$, $\Delta = \omega_{down}\Delta$, *and solve the trust region problem with the new radius to obtain a new trial point. If the trust region radius was just expanded, set* $z = x_t^{old}$.

   (c) *If* $\mu_0 \le ared/pred < \mu_{low}$, *then set* $z = x_t$ *and* $\Delta = \omega_{down}\Delta$.

   (d) *If* $\mu_{low} \le ared/pred \le \mu_{high}$, *set* $z = x_t$.

(e) *If* $\mu_{high} < ared/pred$ *and* $\|s_t\| = \Delta \le C_T \|\nabla f(x_c)\|$, *then set* $z = x_c$, $\Delta = \omega_{up}\Delta$, *and solve the trust region problem with the new radius to obtain a new trial point. Store the old trial point as* x_t^{old} *in case the expansion fails.*

3. $x_+ = z$.

The loop in Algorithm `trtest` serves the same purpose as the loop in a line search algorithm such as Algorithm `steep`. One must design the solution to the trust region problem in such a way that that loop will terminate after finitely many iterations and a general way to do that is the subject of the next section.

We incorporate Algorithm `trtest` into a general trust region algorithm paradigm that we will use for the remainder of this section.

ALGORITHM 3.3.3. `trgen`(x, f)

1. *Initialize* Δ

2. *Do forever*

   (a) *Let* $x_c = x$. *Compute* $\nabla f(x_c)$ *and an approximate Hessian* H_c.

   (b) *Solve the trust region problem to obtain a trial point* x_t.

   (c) *Call* `trtest`(x_c, x_t, x, f, Δ)

Hessians and gradients are computed only in step 2a of Algorithm `trgen`.

3.3.2 Global Convergence of Trust Region Algorithms

While one can, in principal, solve the trust region problem exactly (see §3.3.4), it is simpler and more efficient to solve the problem approximately. It is amazing that one need not do a very good job with the trust region problem in order to get global (and even locally superlinear) convergence.

Our demands of our solutions of the trust region problem and our local quadratic models are modest and readily verifiable. The parameter σ in part 1 of Assumption 3.3.1, like the parameter C_T in (3.29), is used in the analysis but plays no role in implementation. In the specific algorithms that we discuss in this book, σ can be computed. Part 2 follows from well-conditioned and bounded model Hessians if Algorithm `trtest` is used to manage the trust region.

ASSUMPTION 3.3.1.

1. *There is* $\sigma > 0$ *such that*

$$pred = f(x_c) - m_c(x_t) \ge \sigma \|\nabla f(x_c)\| \min(\|s_t\|, \|\nabla f(x_c)\|). \tag{3.30}$$

2. *There is* $M > 0$ *such that either* $\|s_t\| \ge \|\nabla f(x_c)\|/M$ *or* $\|s_t\| = \Delta_c$.

The global convergence theorem based on this assumption should be compared with the similar result on line search methods—Theorem 3.2.4.

THEOREM 3.3.1. *Let* ∇f *be Lipschitz continuous with Lipschitz constant L. Let* $\{x_k\}$ *be generated by Algorithm* `trgen` *and let the solutions for the trust region problems satisfy Assumption* 3.3.1. *Assume that the matrices* $\{H_k\}$ *are bounded. Then either f is unbounded from below,* $\nabla f(x_k) = 0$ *for some finite k, or*

$$\lim_{k \to \infty} \nabla f(x_k) = 0. \tag{3.31}$$

Proof. Assume that $\nabla f(x_k) \neq 0$ for all k and that f is bounded from below. We will show that there is $M_T \in (0,1]$ such that once an iterate is taken (i.e., the step is accepted and the trust region radius is no longer a candidate for expansion), then

$$\|s_k\| \geq M_T \|\nabla f(x_k)\|. \tag{3.32}$$

Assume (3.32) for the present. Since s_k is an acceptable step, Algorithm `trtest` and part 1 of Assumption 3.3.1 imply that

$$ared_k \geq \mu_0 pred_k \geq \mu_0 \|\nabla f(x_k)\| \sigma \min(\|s_k\|, \|\nabla f(x_k)\|).$$

We may then use (3.32) to obtain

$$ared_k \geq \mu_0 \sigma M_T \|\nabla f(x_k)\|^2. \tag{3.33}$$

Now since $f(x_k)$ is a decreasing sequence and f is bounded from below, $\lim_{k\to\infty} ared_k = 0$. Hence (3.33) implies (3.31).

It remains to prove (3.32). To begin note that if $\|s_k\| < \Delta_k$ then by part 2 of Assumption 3.3.1

$$\|s_k\| \geq \|\nabla f(x_k)\|/M.$$

Hence, we need only consider the case in which

$$\|s_k\| = \Delta_k \text{ and } \|s_k\| < \|\nabla f(x_k)\|, \tag{3.34}$$

since if (3.34) does not hold then (3.32) holds with $M_T = \min(1, 1/M)$.

We will complete the proof by showing that if (3.34) holds and s_k is accepted, then

$$\|s_k\| = \Delta_k \geq \frac{2\sigma \min(1-\mu_{high}, (1-\mu_0)\omega_{up}^{-2})}{M+L} \|\nabla f(x_k)\|. \tag{3.35}$$

This will complete the proof with

$$M_T = \min\left(1, 1/M, \frac{2\sigma \min(1-\mu_{high}, (1-\mu_0)\omega_{up}^{-2})}{M+L}\right).$$

Now increase the constant $M > 0$ in part 1 of Assumption 3.3.1 if needed so that

$$\|H_k\| \leq M \text{ for all } k. \tag{3.36}$$

We prove (3.35) by showing that if (3.34) holds and (3.35) does not hold for a trial step s_t, then the trust region radius will be expanded and the step corresponding to the larger radius will be acceptable. Let s_t be a trial step such that $\|s_t\| < \|\nabla f(x_k)\|$ and

$$\|s_t\| = \Delta_t < \frac{2\sigma \min(1-\mu_{high}, (1-\mu_0)\omega_{up}^{-2})}{M+L} \|\nabla f(x_k)\|. \tag{3.37}$$

We use the Lipschitz continuity of ∇f and (3.36) to obtain

$$\begin{aligned} ared_t &= -\nabla f(x_k)^T s_t - \int_0^1 (\nabla f(x_k + ts_t) - \nabla f(x_k))^T s_t \, dt \\ &= pred_t + s_t^T H_k s_t/2 - \int_0^1 (\nabla f(x_k + ts_t) - \nabla f(x_k))^T s_t \, dt \\ &\geq pred_t - (M+L)\|s_t\|^2/2. \end{aligned}$$

Therefore, using (3.30) from Assumption 3.3.1, we have

$$
\begin{aligned}
\frac{ared_t}{pred_t} &\geq 1 - \frac{(M+L)\|s_t\|^2}{2pred_t} \\
&\geq 1 - \frac{(M+L)\|s_t\|^2}{2\sigma\|\nabla f(x_k)\| \min(\|\nabla f(x_k)\|, \|s_t\|)}.
\end{aligned} \tag{3.38}
$$

Now since $\|s_t\| < \|\nabla f(x_k)\|$ by (3.34) we have

$$
\min(\|\nabla f(x_k)\|, \|s_t\|) = \|s_t\|
$$

and hence

$$
\frac{ared_k}{pred_k} \geq 1 - \frac{(M+L)\|s_t\|}{2\|\nabla f(x_k)\|\sigma} > \mu_{high} \tag{3.39}
$$

by (3.37). Hence, an expansion step will be taken by replacing Δ_t by $\Delta_t^+ = \omega_{up}\Delta_t$ and s_t by s_t^+, the minimum of the quadratic model in the new trust region.

Now (3.38) still holds and, after the expansion,

$$
\|s_t^+\| \leq \omega_{up}\|s_t\| < \omega_{up}\|\nabla f(x_k)\|.
$$

So

$$
\min(\|\nabla f(x_k)\|, \|s_t^+\|) > \|s_t^+\|/\omega_{up}.
$$

Hence,

$$
\begin{aligned}
\frac{ared_t^+}{pred_t^+} &\geq 1 - \frac{(M+L)\|s_t^+\|^2}{2\sigma\|\nabla f(x_k)\| \min(\|\nabla f(x_k)\|, \|s_t^+\|)} \\
&\geq 1 - \frac{(M+L)\omega_{up}\|s_t^+\|}{2\|\nabla f(x_k)\|\sigma} \geq 1 - \frac{(M+L)\omega_{up}^2\|s_t\|}{2\|\nabla f(x_k)\|\sigma} \geq \mu_0
\end{aligned}
$$

by (3.37). Hence, the expansion will produce an acceptable step. This means that if the final accepted step satisfies (3.34), it must also satisfy (3.35). This completes the proof. □

3.3.3 A Unidirectional Trust Region Algorithm

The most direct way to compute a trial point that satisfies Assumption 3.3.1 is to mimic the line search and simply minimize the quadratic model in the steepest descent direction subject to the trust region bound constraints.

In this algorithm, given a current point x_c and trust region radius Δ_c, our trial point is the minimizer of

$$
\psi_c(\lambda) = m_c(x_c - \lambda\nabla f(x_c))
$$

subject to the constraint that

$$
x(\lambda) = x_c - \lambda\nabla f(x_c) \in \mathcal{T}(\Delta_c).
$$

Clearly the solution is $x(\hat{\lambda})$ where

$$
\hat{\lambda} = \begin{cases} \frac{\Delta_c}{\|\nabla f(x_c)\|} & \text{if } \nabla f(x_c)^T H_c \nabla f(x_c) \leq 0\,, \\ \min\left(\frac{\|\nabla f(x_c)\|^2}{\nabla f(x_c)^T H_c \nabla f(x_c)}, \frac{\Delta_c}{\|\nabla f(x_c)\|}\right) & \text{if } \nabla f(x_c)^T H_c \nabla f(x_c) > 0. \end{cases} \tag{3.40}
$$

$x(\hat{\lambda})$, the minimizer of the quadratic model in the steepest descent direction, subject to the trust region bounds, is called the *Cauchy point*. We will denote the Cauchy point by x_c^{CP}.[1]

Then with x^{CP} as trial point, one can use Theorem 3.3.1 to derive a global convergence theorem for the unidirectional trust region.

THEOREM 3.3.2. *Let ∇f be Lipschitz continuous with Lipschitz constant L. Let $\{x_k\}$ be generated by Algorithm* `trgen` *with $x_t = x^{CP}$ and* (3.40). *Assume that the matrices $\{H_k\}$ are bounded. Then either $f(x_k)$ is unbounded from below, $\nabla f(x_k) = 0$ for some finite k, or*

$$\lim_{k\to\infty} \nabla f(x_k) = 0.$$

Proof. We show that x_t satisfies part 2 of Assumption 3.3.1. If $\|s_t\| = \Delta_c$ then the assertion holds trivially. If $\|s_t\| < \Delta_c$ then, by definition of x_c^{CP},

$$s_t = -\frac{\|\nabla f(x_c)\|^2 \nabla f(x_c)}{\nabla f(x_c)^T H_c \nabla f(x_c)}.$$

Hence, if $\|H_c\| \le M$,

$$\|s_t\| \ge \|\nabla f(x_c)\|/M$$

as asserted.

We leave the proof that x_t satisfies part 1 for the reader (exercise 3.5.8). □

The assumptions we used are stronger that those in, for example, [104] and [223], where

$$\liminf \|\nabla f(x_k)\| = 0$$

rather than $\nabla f(x_k) \to 0$ is proved.

3.3.4 The Exact Solution of the Trust Region Problem

The theory of constrained optimization [117], [104] leads to a characterization of the solutions of the trust region problem. In this section we derive that characterization via an elementary argument (see also [84], [242], and [109]). This book focuses on approximate solutions, but the reader should be aware that the exact solution can be computed accurately [192], [243].

THEOREM 3.3.3. *Let $g \in R^N$ and let A be a symmetric $N \times N$ matrix. Let*

$$m(s) = g^T s + s^T A s/2.$$

A vector s is a solution to

$$\min_{\|s\|\le\Delta} m(s) \tag{3.41}$$

if and only if there is $\nu \ge 0$ such that

$$(A + \nu I)s = -g$$

and either $\nu = 0$ or $\|s\| = \Delta$.

Proof. If $\|s\| < \Delta$ then $\nabla m(s) = g + As = 0$, and the conclusion follows with $\nu = 0$. To consider the case where $\|s\| = \Delta$, let $\lambda_1 \le \lambda_2 \le \cdots \lambda_N$ be the eigenvalues of A.

---

[1] In some of the literature, [84], for example, H_c is assumed to be positive definite and the Cauchy point is taken to be the global minimizer of the quadratic model.

Clearly, for any ν,

$$\begin{aligned} m(s) &= g^T s + s^T A s/2 \\ &= g^T s + s^T (A + \nu I) s/2 - \nu \Delta^2/2. \end{aligned}$$

Consider the function, defined for $\nu > \nu_0 = \max(0, -\lambda_1)$,

$$s(\nu) = -(A + \nu I)^{-1} g.$$

Since

$$\lim_{\nu \to \infty} s(\nu) = 0$$

and $\|s(\nu)\|$ is a continuous decreasing function of $\nu \in (\nu_0, \infty)$ we see that if

$$\lim_{\nu \to \nu_0} \|s(\nu)\| > \Delta$$

then there is a unique ν such that $\|s(\nu)\| = \Delta$. Since $\nu \geq \nu_0$, $A + \nu I$ is positive semidefinite; therefore, $s(\nu)$ is a global minimizer of

$$g^T s + s^T (A + \nu I) s/2.$$

Hence, we must have

$$m(s) \geq m(s(\nu))$$

for all s such that $\|s\| = \Delta$. Hence, $s(\nu)$ is a solution of (3.41).

The remaining case is

$$\lim_{\nu \to \nu_0} \|s(\nu)\| \leq \Delta.$$

This implies that g is orthogonal to the nontrivial space $\mathcal{S}_0$ of eigenfunctions corresponding to $-\nu_0$ (for otherwise the limit would be infinite). If we let $s = s_1 + s_2$, where s_2 is the projection of s onto $\mathcal{S}_0$, we have

$$\begin{aligned} m(s) &= s_1^T g + s_1^T (A + \nu_0) s_1/2 + s_2^T (A + \nu_0) s_2/2 - \nu_0 \Delta^2/2 \\ &= s_1^T g + s_1^T (A + \nu_0) s_1/2 - \nu_0 \Delta^2/2. \end{aligned}$$

Hence, $m(s)$ is minimized by setting s_1 equal to the minimum norm solution of $(A+\nu_0)x = -g$ (which exists by orthogonality of g to $\mathcal{S}_0$) and letting s_2 be any element of $\mathcal{S}_0$ such that

$$\|s_2\|^2 = \Delta^2 - \|s_1\|^2.$$

This completes the proof. □

3.3.5 The Levenberg–Marquardt Parameter

The solution of the trust region problem presented in §3.3.4 suggests that, rather than controlling Δ, one could set

$$s_t = -(\nu I + H_c)^{-1} g,$$

adjust ν in response to $ared/pred$ instead of Δ, and still maintain global convergence. A natural application of this idea is control of the Levenberg–Marquardt parameter. This results in a much simpler algorithm than Levenberg–Marquardt–Armijo in that the stepsize control can be eliminated. We need only vary the Levenberg–Marquardt parameter as the iteration progresses.

We present the algorithm from [190] to illustrate this point. For an inexact formulation, see [276].

The Levenberg–Marquardt quadratic model of least squares objective

$$f(x) = \frac{1}{2}\sum_{i=1}^{M} \|r_i(x)\|_2^2 = \frac{1}{2}R(x)^T R(x)$$

with parameter ν_c at the point x_c is

$$(3.42) \qquad \begin{aligned} m_c(x) &= f(x_c) + (x - x_c)^T R'(x_c)^T R(x_c) \\ &+ \tfrac{1}{2}(x - x_c)^T (R'(x_c)^T R'(x_c) + \nu_c I)(x - x_c). \end{aligned}$$

The minimizer of the quadratic model is the trial point

$$(3.43) \qquad x_t = x_c - (R'(x_c)^T R'(x_c) + \nu_c I)^{-1} R'(x_c)^T R(x_c),$$

the step is $s = x_t - x_c$, and the predicted reduction is

$$\begin{aligned} pred &= m(x_c) - m(x_t) = -s^T R'(x_c)^T R(x_c) - \tfrac{1}{2}s^T (R'(x_c)^T R'(x_c) + \nu_c I)s \\ &= -s^T R'(x_c)^T R(x_c) + \tfrac{1}{2}s^T R'(x_c)^T R(x_c) = -\tfrac{1}{2}s^T \nabla f(x_c). \end{aligned}$$

The algorithm we present below follows the trust region paradigm and decides on accepting the trial point and on adjustments in the Levenberg–Marquardt parameter by examinaing the ratio

$$\begin{aligned} \frac{ared}{pred} &= \frac{f(x_c) - f(x_t)}{m(x_c) - m(x_t)} \\ &= -2\frac{f(x_c) - f(x_t)}{s^T \nabla f(x_c)}. \end{aligned}$$

In addition to the trust region parameters $0 < \omega_{down} < 1 < \omega_{up}$ and $\mu_0 \le \mu_{low} < \mu_{high}$ we require a default value ν_0 of the Levenberg–Marquardt parameter.

The algorithm for testing the trial point differs from Algorithm `trtest` in that we decrease (increase) ν rather that increasing (decreasing) a trust region radius if $ared/pred$ is large (small). We also attempt to set the Levenberg–Marquardt parameter to zero when possible in order to recover the Gauss–Newton iteration's fast convergence for small residual problems.

ALGORITHM 3.3.4. `trtestlm`(x_c, x_t, x_+, f, ν)

1. $z = x_c$

2. *Do while* $z = x_c$

   (a) $ared = f(x_c) - f(x_t)$, $s_t = x_t - x_c$, $pred = -\nabla f(x_c)^T s_t/2$.

   (b) *If* $ared/pred < \mu_0$ *then set* $z = x_c$, $\nu = \max(\omega_{up}\nu, \nu_0)$, *and recompute the trial point with the new value of* ν.

   (c) *If* $\mu_0 \le ared/pred < \mu_{low}$, *then set* $z = x_t$ *and* $\nu = \max(\omega_{up}\nu, \nu_0)$.

   (d) *If* $\mu_{low} \le ared/pred$, *then set* $z = x_t$.
   If $\mu_{high} < ared/pred$, *then set* $\nu = \omega_{down}\nu$.
   If $\nu < \nu_0$, *then set* $\nu = 0$.

3. $x_+ = z$.

The Levenberg–Marquardt version of Algorithm `trgen` is simple to describe and implement.

ALGORITHM 3.3.5. `levmar`$(x, R, kmax)$

1. *Set* $\nu = \nu_0$.
2. *For* $k = 1, \dots, kmax$
   (a) *Let* $x_c = x$.
   (b) *Compute* R, f, R', *and* ∇f; *test for termination.*
   (c) *Compute* x_t *using* (3.43).
   (d) *Call* `trtestlm`(x_c, x_t, x, f, ν)

We state a convergence result [190], [276] without proof.

THEOREM 3.3.4. *Let R be Lipschitz continuously differentiable. Let $\{x_k\}$ and $\{\nu_k\}$ be the sequence of iterates and Levenberg–Marquardt parameters generated by Algorithm* `levmar` *with $kmax = \infty$. Assume that $\{\nu_k\}$ is bounded from above. Then either $R'(x_k)^T R(x_k) = 0$ for some finite k or*

$$\lim_{k\to\infty} R'(x_k)^T R(x_k) = 0.$$

Moreover, if x^* is a limit point of $\{x_k\}$ for which $R(x^*) = 0$ and $R'(x^*)$ has full rank, then $x_k \to x^*$ q-quadratically and $\nu_k = 0$ for k sufficiently large.

3.3.6 Superlinear Convergence: The Dogleg

The convergence of the unidirectional trust region iteration can be as slow as that for steepest descent. To improve the convergence speed in the terminal phase we must allow for approximations to the Newton direction. The power of trust region methods is the ease with which the transition from steepest descent, with its good global properties, to Newton's method can be managed.

We define the *Newton point* at x_c as

$$x_c^N = x_c - H_c^{-1}\nabla f(x_c).$$

If H_c is spd, the Newton point is the global minimizer of the local quadratic model. On the other hand, if H_c has directions of negative curvature the local quadratic model will not have a finite minimizer, but the Newton point is still useful. Note that if $H = I$ the Newton point and the Cauchy point are the same if the Newton point is inside the trust region.

We will restrict our attention to a special class of algorithms that approximate the solution of the trust region problem by minimizing m_c along a piecewise linear path $\mathcal{S} \subset \mathcal{T}(\Delta)$. These paths are sometimes called *doglegs* because of the shapes of the early examples [84], [80], [218], [217], [220]. In the case where $\nabla^2 f(x)$ is spd, one may think of the dogleg path as a piecewise linear approximation to the path with parametric representation

$$\{x - (\lambda I + \nabla^2 f(x))^{-1}\nabla f(x) \,|\, 0 \le \lambda\}.$$

This is the path on which the exact solution of the trust region problem lies.

The next step up from the unidirectional path, the *classical dogleg* path [220], has as many as three nodes, x_c, x_c^{CP*}, and x_c^N. Here x_c^{CP*} is the global minimizer of the quadratic model in the steepest descent direction, which will exist if and only if $\nabla f(x_c)^T H_c \nabla f(x_c) > 0$. If x_c^{CP*} exists and

$$(x_c^N - x_c^{CP*})^T(x_c^{CP*} - x_c) > 0, \tag{3.44}$$

we will let x_c^N be the terminal node. If (3.44) holds, as it always will if H_c is spd, then the path can be parameterized by the distance from x_c and, moreover, m_c decreases along the path. If (3.44) does not hold, we do not use x_c^N as a node and revert to the unidirectional path in the steepest descent direction.

Note that (3.44) implies

$$\nabla f(x_c)^T(x_c^N - x_c) < 0. \tag{3.45}$$

We can express the conditions for using the three node path rather than the unidirectional path very simply. If x_c^{CP} is on the boundary of the trust region then we accept x_c^{CP} as the trial point. If $x_c^{CP} = x_c^{CP*}$ is in the interior of the trust region, then we test (3.44) to decide what to do.

With this in mind our trial point for the classical dogleg algorithm will be

$$x^D(\Delta) = \begin{cases} x^{CP} & \text{if } \|x_c - x_c^{CP}\| = \Delta \\ & \text{or } x^{CP*} \text{ exists and (3.44) fails,} \\ x^N & \text{if } \|x_c - x_c^{CP}\| < \|x_c - x_c^N\| \le \Delta \\ & \text{and (3.44) holds,} \\ y^D(\Delta) & \text{otherwise.} \end{cases} \tag{3.46}$$

Here $y^D(\Delta)$ is the unique point between x_c^{CP} and x_c^N such that $\|x^D - x_c\| = \Delta$.

The important properties of dogleg methods are as follows:

- No two points on the path have the same distance from x_c; hence the path may be parameterized as $x(s)$, where $s = \|x(s) - x_c\|$.
- $m_c(x(s))$ is a strictly decreasing function of s.

This enables us to show that the dogleg approximate solution of the trust region problem satisfies Assumption 3.3.1 and apply Theorem 3.3.1 to conclude global convergence. Superlinear convergence will follow if H_k is a sufficiently good approximation to $\nabla^2 f(x_k)$.

LEMMA 3.3.5. *Let x_c, H_c, and Δ be given. Let H_c be nonsingular,*

$$s^N = -H_c^{-1}\nabla f(x_c), \textit{ and } x^N = x_c + s^N.$$

Assume that $\nabla f(x_c)^T H_c \nabla f(x_c) > 0$ and let

$$s^{CP*} = x^{CP*} - x_c = -\frac{\|\nabla f(x_c)\|^2}{\nabla f(x_c)^T H_c \nabla f(x_c)} \nabla f(x_c).$$

Let S be the piecewise linear path from x_c to x^{CP*} to x^N. Then if

$$(s^N - s^{CP*})^T s^{CP*} > 0, \tag{3.47}$$

for any $\delta \le \|s^N\|$ there is a unique point $x(\delta)$ on S such that

$$\|x(\delta) - x_c\| = \delta.$$

Proof. Clearly the statement of the result holds on the segment of the path from x to x^{CP*}. To prove the result on the segment from x^{CP*} to x^N we must show that

$$\phi(\lambda) = \frac{1}{2}\|(1-\lambda)s^{CP*} + \lambda s^N\|^2$$

is strictly monotone increasing for $\lambda \in (0, 1)$.

Since (3.47) implies that

$$\|s^N\|\|s^{CP*}\| \geq (s^N)^T s^{CP*} > \|s^{CP*}\|^2$$

and therefore that $\|s^N\| > \|s^{CP*}\|$, we have

$$\begin{aligned} \phi'(\lambda) &= (s^N - s^{CP*})^T((1-\lambda)s^{CP*} + \lambda s^N) \\ &= -(1-\lambda)\|s^{CP*}\|^2 + (1-\lambda)(s^N)^T s^{CP*} + \lambda\|s^N\|^2 - \lambda(s^N)^T s^{CP*} \\ &> \lambda(\|s^N\|^2 - (s^N)^T s^{CP*}) \geq \lambda(\|s^N\| - \|s^{CP*}\|)\|s^N\| > 0. \end{aligned}$$

Hence, ϕ is an increasing function and the proof is complete. □

The next stage is to show that the local quadratic model decreases on the dogleg path $\mathcal{S}$.

LEMMA 3.3.6. *Let the assumptions of Lemma 3.3.5 hold. Then the local quadratic model*

$$m_c(x) = f(x_c) + \nabla f(x_c)^T(x - x_c) + \frac{1}{2}(x - x_c)^T H_c(x - x_c)$$

is strictly monotone decreasing on $\mathcal{S}$.

Proof. Since x_c^{CP*} is the minimum of the local quadratic model in the steepest descent direction, we need only show that m_c is strictly decreasing on the segment of the path between x_c^{CP*} and x^N. Set

$$\begin{aligned} \psi(\lambda) &= m_c(x_c + (1-\lambda)s^{CP*} + \lambda s^N) \\ &= f(x_c) + \nabla f(x_c)^T((1-\lambda)s^{CP*} + \lambda s^N) \\ &\quad + \tfrac{1}{2}((1-\lambda)s^{CP*} + \lambda s^N)^T H_c((1-\lambda)s^{CP*} + \lambda s^N). \end{aligned}$$

Noting that $H_c s^N = -\nabla f(x_c)$ and $s^{CP*} = -\hat{\lambda}\nabla f(x_c)$, we obtain

$$\begin{aligned} \psi(\lambda) &= f(x_c) - \hat{\lambda}(1-\lambda)^2\|\nabla f(x_c)\|^2 \\ &\quad + \lambda(1-\lambda/2)\nabla f(x_c)^T s^N \\ &\quad + \tfrac{1}{2}(1-\lambda)^2\hat{\lambda}^2\nabla f(x_c)^T H_c \nabla f(x_c). \end{aligned}$$

Therefore,

$$\begin{aligned} \psi'(\lambda) &= 2\hat{\lambda}(1-\lambda)\|\nabla f(x_c)\|^2 \\ &\quad + (1-\lambda)\nabla f(x_c)^T s^N - (1-\lambda)\hat{\lambda}^2\nabla f(x_c)^T H_c \nabla f(c_c). \end{aligned}$$

Since

$$\hat{\lambda}\nabla f(x_c)^T H_c \nabla f(c_c) = \|\nabla f(x_c)\|^2$$

we have, using (3.44),

$$\begin{aligned} \psi'(\lambda) &= (1-\lambda)(\hat{\lambda}\|\nabla f(x_c)\|^2 - \nabla f(x_c)^T H_c^{-1}\nabla f(x_c)) \\ &= (1-\lambda)\nabla f(x_c)^T(\hat{\lambda}\nabla f(x_c) - H_c^{-1}\nabla f(x_c)) \\ &= \frac{1-\lambda}{\hat{\lambda}}(x_c - x_c^{CP*})^T(x_c^N - x_c) < 0, \end{aligned}$$

completing the proof. □

At this point we have shown that the approximate trust region problem has a unique solution. To prove global convergence we need only verify that the approximate solution of the trust region problem x^D satisfies Assumption 3.3.1.

THEOREM 3.3.7. *Let ∇f be Lipschitz continuous with Lipschitz constant L. Let $\{x_k\}$ be generated by Algorithm* `trgen` *and the solutions for the trust region problem be given by (3.46). Assume that the matrices $\{H_k\}$ are bounded. Then either $f(x_k)$ is unbounded from below, $\nabla f(x_k) = 0$ for some finite k, or*

$$\lim_{k \to \infty} \nabla f(x_k) = 0. \tag{3.48}$$

Proof. We need to check that the solutions of the trust region problem satisfy Assumption 3.3.1. Part 2 of the assumption follows from the definition, (3.46), of x^D and the boundedness of the approximate Hessians. Let

$$\|H_k\| \leq M$$

for all k. If $\|s_k\| < \Delta$, then (3.46) implies that (3.44) must hold and so $x_t = x_k^N$ is the Newton point. Hence,

$$\|s_k\| = \|x_k - x_k^N\| = \|H_k^{-1} \nabla f(x_k)\| \geq \|\nabla f(x_k)\|/M,$$

which proves part 2.

Verification of part 1 will complete the proof. There are several cases to consider depending on how x^D is computed.

If $x^D = x^{CP}$ then either $\|s^{CP}\| = \Delta_c$ or (3.44) fails. We first consider the case where $\nabla f(x_c)^T H_c \nabla f(x_c) \leq 0$. In that case $\|s^{CP}\| = \Delta_c$ and $\hat{\lambda} = \Delta_c / \|\nabla f(x_c)\|$. Therefore,

$$\begin{aligned} pred &= \hat{\lambda}\|\nabla f(x_c)\|^2 - \frac{\hat{\lambda}^2}{2} \nabla f(x_c)^T H_c \nabla f(x_c) \\ &= \Delta_c \|\nabla f(x_c)\| - \Delta_c^2 \frac{\nabla f(x_c)^T H_c \nabla f(x_c)}{2\|\nabla f(x_c)\|^2} \\ &\geq \Delta_c \|\nabla f(x_c)\| = \|s\| \|\nabla f(x_c)\|. \end{aligned}$$

Hence (3.30) holds with $\sigma = 1$.

Now assume that $\nabla f(x_c)^T H_c \nabla f(x_c) > 0$ and $\|s^{CP}\| = \Delta_c$. In this case

$$\frac{\|\nabla f(x_c)\|^2}{\nabla f(x_c)^T H_c \nabla f(x_c)} \geq \frac{\Delta_c}{\|\nabla f(x_c)\|}$$

and so

$$\begin{aligned} pred &= \hat{\lambda}\|\nabla f(x_c)\|^2 - \frac{\hat{\lambda}^2}{2} \nabla f(x_c)^T H_c \nabla f(x_c) \\ &= \Delta_c \|\nabla f(x_c)\| - \Delta_c^2 \frac{\nabla f(x_c)^T H_c \nabla f(x_c)}{2\|\nabla f(x_c)\|^2} \\ &\geq \Delta_c \|\nabla f(x_c)\|/2, \end{aligned}$$

which is (3.30) with $\sigma = 1/2$.

If (3.44) fails, $\nabla f(x_c)^T H_c \nabla f(x_c) > 0$, and $\|s^{CP}\| < \Delta_c$, then

$$\hat{\lambda} = \frac{\|\nabla f(x_c)\|^2}{\nabla f(x_c)^T H_c \nabla f(x_c)},$$

and

$$\begin{aligned} pred &= \hat{\lambda}\|\nabla f(x_c)\|^2 - \frac{\hat{\lambda}^2}{2}\nabla f(x_c)^T H_c \nabla f(x_c) \\ &= \frac{\|\nabla f(x_c)\|^4}{2\nabla f(x_c)^T H_c \nabla f(x_c)} = \frac{\hat{\lambda}\|\nabla f(x_c)\|^2}{2} \\ &= \frac{\|s\|\|\nabla f(x_c)\|}{2}, \end{aligned}$$

which is (3.30) with $\sigma = 1/2$.

The final case is if (3.44) holds and $x^D \neq x^{CP}$. In that case the predicted reduction is more than *Cauchy decrease*, i.e., the decrease obtained by taking the Cauchy point, and hence

$$\begin{aligned} pred &\geq \frac{\|\nabla f(x_c)\|^4}{2\nabla f(x_c)^T H_c \nabla f(x_c)} \\ &\geq \frac{\|\nabla f(x_c)\|^2}{2M}, \end{aligned}$$

which is (3.30) with $\sigma = 1/(2M)$. This completes the proof. □

The last part of the proof of this theorem is very important, asserting that any solution of the trust region problem for which *pred* is at least a fixed fraction of Cauchy decrease will give global convergence. We refer the reader to [232] and [104] for a more general and detailed treatment using this point of view.

COROLLARY 3.3.8. *Any algorithm for solving the trust region problem that satisfies for some* $\tau > 0$

$$pred \geq \tau(m_c(x_c) - m_c(x_c^{CP}))$$

satisfies (3.30) *for* $\sigma = \tau/2$.

The trust region CG algorithm we present in §3.3.7 can be analyzed with this corollary.

If $H_k = \nabla^2 f(x_k)$ or a sufficiently good approximation, then the classical dogleg will become Newton's method (or a superlinearly convergent method) as the iterations approach a minimizer that satisfies the standard assumptions. Hence, the algorithm makes a smooth and automatic transition into the superlinearly convergent stage.

THEOREM 3.3.9. *Let* ∇f *be Lipschitz continuous with Lipschitz constant L. Let* $\{x_k\}$ *be generated by Algorithm* `trgen` *and the solutions for the trust region problem are given by* (3.46). *Assume that* $H_k = \nabla^2 f(x_k)$ *and that the matrices* $\{H_k\}$ *are bounded. Let f be bounded from below. Let* x^* *be a minimizer of f at which the standard assumptions hold. Then if* x^* *is a limit point of* x_k*, then* $x_k \to x^*$ *and the convergence is locally q-quadratic.*

Proof. Since x^* is a limit point of $\{x_k\}$, there is, for any $\rho > 0$, a k sufficiently large so that

$$\|e_k\| < \rho, \|H_k\| \leq 2\|\nabla^2 f(x^*)\|, \|H_k^{-1}\| \leq 2\|(\nabla^2 f(x^*))^{-1}\|,$$

and x_k is near enough for the assumptions of Theorem 2.3.2 to hold. If H_k is spd, so is H_k^{-1} and for such k, (3.44) holds. Hence, the dogleg path has the nodes x_k, x_k^{CP}, and x_k^N. Moreover, if ρ is sufficiently small, then

$$\|H_k^{-1}\nabla f(x_k)\| \leq 2\|e_k\| \leq 2\rho.$$

We complete the proof by showing that if ρ is sufficiently small, the trust region radius will be expanded if necessary until the Newton step is in the trust region. Once we do this, the proof is complete as then the local quadratic convergence of Newton's method will take over.

Now

$$pred_k \geq \|s_k\| \|\nabla f(x_k)\|/2$$

by the proof of Theorem 3.3.7. Using $H_k = \nabla^2 f(x_k)$ we have

$$\begin{aligned} ared_k &= -\nabla f(x_k)^T s_{tk} - \int_0^1 (\nabla f(x_k + t s_{tk}) - \nabla f(x_k))^T s_{tk}\, dt \\ &= pred_k + s_{tk}^T \nabla^2 f(x_k) s_{tk}/2 - \int_0^1 (\nabla f(x_k + t s_{tk}) - \nabla f(x_k))^T s_{tk}\, dt \\ &= pred_k + O(\|s_k\| \|\nabla f(x_k)\| \rho) \end{aligned}$$

and therefore $ared/pred = 1 - O(\rho)$. Hence, for ρ sufficiently small, the trust region radius will be increased, if necessary, until the Newton point is inside the trust region and then a Newton step will be taken. This completes the proof. □

The classical dogleg algorithm is implemented in Algorithm `ntrust`, which uses the trust radius adjustment scheme from Algorithm `trtest`. It is to be understood that `trtest` is implemented so that x_t is given by (3.46) and hence `trtest` only samples points on the piecewise linear search path determined by the Cauchy point, the Newton point, and (3.44).

ALGORITHM 3.3.6. $\texttt{ntrust}(x, f, \tau)$

1. *Compute* $f(x)$ *and* $\nabla f(x)$
2. $\tau = \tau_a + \tau_r \|\nabla f(x)\|$
3. *Do while* $\|\nabla f(x)\| > \tau$
   (a) *Compute and factor* $\nabla^2 f(x)$
   (b) *Compute the Cauchy and Newton points and test* (3.44)
   (c) *Call* $\texttt{trtest}(x, x_t, x_+, f, \Delta)$
   (d) *Compute* $f(x_+)$ *and* $\nabla f(x_+)$; $x = x_+$

We implement Algorithm `ntrust` in the collection of MATLAB codes.

3.3.7 A Trust Region Method for Newton–CG

In this section we present a brief account of an algorithm from [247] (see also [257]) that combines the trust region paradigm of §3.3.6 with the inexact Newton ideas of §2.5.2. We follow §2.5.2 and denote the preconditioner by M and let $C = M^{-1}$. We solve the scaled trust region problem

$$\min_{\|d\|_C \leq \Delta} \phi(d),$$

where the quadratic model is still

$$\phi(d) = \nabla f(x)^T d + \frac{1}{2} d^T \nabla^2 f(x) d.$$

Here the C-norm is

$$\|d\|_C = (d^T C d)^{1/2}.$$

The algorithmic description of the trust region problem solver from the TR–CG method given below is from [162]. In [247] the algorithm is expressed in terms of C rather than M.

This is a dogleg method in that the approximate solution of the trust region problem lies on a piecewise linear path with the CG iterations as nodes. As long as CG is performing properly (i.e., $p^T w > 0$) nodes are added to the path until the path intersects the trust region boundary. If a direction of indefiniteness is found ($p^T w \leq 0$), then that direction is followed to the boundary. In this way a negative curvature direction, if found in the course of the CG iteration, can be exploited.

The inputs to Algorithm `trcg` are the current point x, the objective f, the forcing term η, and the current trust region radius Δ. The output is the approximate solution of the trust region problem d. This algorithm is not the whole story, as once the trust region problem is solved approximately, one must use $f(x_c + d)$ to compute $ared$ and then make a decision on how the trust region radius should be changed. Our formulation differs from that in [247] in that the termination criterion measures relative residuals in the l^2-norm rather than in the C-norm. This change in the norm has no effect on the analysis in [247], and, therefore, we can apply the results in §2.5 directly to draw conclusions about local convergence.

ALGORITHM 3.3.7. `trcg`$(d, x, f, M, \eta, \Delta, kmax)$

1. $r = -\nabla f(x)$, $\rho_0 = \|r\|_2^2$, $k = 1$, $d = 0$
2. *Do While* $\sqrt{\rho_{k-1}} > \eta \|\nabla f(x)\|_2$ *and* $k < kmax$
   (a) $z = Mr$
   (b) $\tau_{k-1} = z^T r$
   (c) *if* $k = 1$ *then* $\beta = 0$ *and* $p = z$
   else
   $\beta = \tau_{k-1}/\tau_{k-2}$, $p = z + \beta p$
   (d) $w = \nabla^2 f(x) p$
   If $p^T w \leq 0$ *then*
   Find τ *such that* $\|d + \tau p\|_C = \Delta$
   $d = d + \tau p$; *return*
   (e) $\alpha = \tau_{k-1}/p^T w$
   (f) $r = r - \alpha w$
   (g) $\rho_k = r^T r$
   (h) $\hat{d} = d + \alpha p$
   (i) *If* $\|\hat{d}\|_C > \Delta$ *then*
   Find τ *such that* $\|d + \tau p\|_C = \Delta$
   $d = d + \tau p$; *return*
   (j) $d = \hat{d}$; $k = k + 1$

Algorithm `trcg` does what we would expect a dogleg algorithm to do in that the piecewise linear path determined by the iteration moves monotonically away from x (in the $\|\cdot\|_C$-norm!) and the quadratic model decreases on that path [247]. Algorithm `trcg` will, therefore, compute the same Newton step as Algorithm `fdpcg`. One might think that it may be difficult to compute the C-norm if one has, for example, a way to compute the action of M on a vector that does not require computation of the matrix C. However, at the cost of storing two additional vectors we can update Cp and Cd as the iteration progresses. So, when p is updated to $z + \beta p$ then $Cp = r + \beta Cp$ can be updated at the same time without computing the product of C with p. Then $\|p\|_C = p^T Cp$. Similarly $d = d + \tau p$ implies that $Cd = Cd + \tau Cp$.

Algorithm `cgtrust` combines the solution of the trust region problem from `trcg`, the trust region radius adjustment scheme from `trtest`, and (indirectly) the locally convergent algorithm `newtcg`. The result fits nicely into our paradigm algorithm `trgen`.

ALGORITHM 3.3.8. `cgtrust`(x, f, τ)

1. *Initialize* Δ, M, η, $kmax$.
2. *Do forever*
   (a) *Let* $x_c = x$. *Compute* $\nabla f(x_c)$.
   (b) *Call* `trcg`$(d, x, f, M, \eta, \Delta, kmax)$ *to solve the trust region subproblem.* *Set* $x_t = x + d$.
   (c) *Call* `trtest`(x_c, x_t, x, f, Δ), *solving the trust region subproblem with Algorithm* `trcg`.
   (d) *Update* η.

Theorem 3.3.10 combines several results from [247].

THEOREM 3.3.10. *Let* f *be twice Lipschitz continuously differentiable. Let* M *be a given positive definite matrix and let* $\{\eta_n\}$ *satisfy* $0 < \eta_n < 1$ *for all* n. *Let* $\{x_n\}$ *be the sequence generated by Algorithm* `cgtrust` *and assume that* $\{\|\nabla^2 f(x_n)\|\}$ *is bounded. Then*

$$\lim_{n \to \infty} \nabla f(x_n) = 0. \tag{3.49}$$

Moreover, if x^* *is a local minimizer for which the standard assumptions hold and* $x_n \to x^*$, *then*

- *if* $\eta_n \to 0$ *the convergence is q-superlinear, and*
- *if* $\eta_n \leq K_\eta \|\nabla f(x_n)\|^p$ *for some* $K_\eta > 0$ *the convergence is q-superlinear with q-order* $1 + p$.

Finally, there are δ *and* Δ *such that if* $\|x_0 - x^*\| \leq \delta$ *and* $\Delta_0 \leq \Delta$ *then* $x_n \to x^*$.

One can, as we do in the MATLAB code `cgtrust`, replace the Hessian–vector product with a difference Hessian. The accuracy of the difference Hessian and the loss of symmetry present the potential problem that was mentioned in §2.5. Another, very different, approach is to approximate the exact solution of the trust region subproblem with an iterative method [243].

3.4 Examples

The results we report here used the MATLAB implementations of steepest descent, `steep.m`, damped Gauss–Newton, `gaussn.m`, the dogleg trust region algorithm for Newton's method, `ntrust.m`, and the PCG–dogleg algorithms, `cgtrust.m`, from the software collection.

Our MATLAB implementation of Algorithm `steep` guards against extremely poor scaling and very long steps by setting λ to

$$\lambda_0 = \min(1, 100/(1 + \|\nabla f(x)\|)) \tag{3.50}$$

at the beginning of the line search. We invite the reader in Exercise 3.5.3 to attempt the control example with $\lambda_0 = 1$.

We not only present plots, which are an efficient way to understand convergence rates, but we also report counts of function, gradient, and Hessian evaluations and the results of the MATLAB `flops` command.

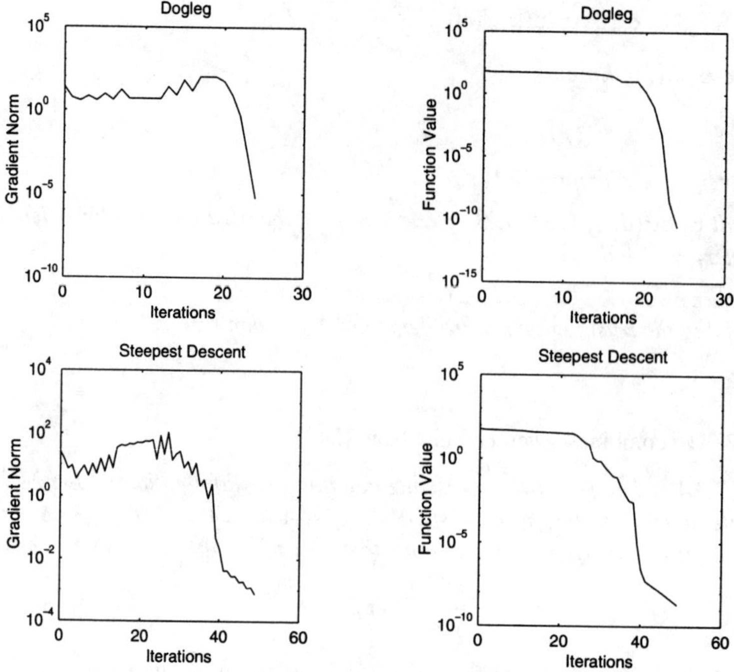

Figure 3.1: *Steepest Descent and Newton–Dogleg for Parameter ID Problem*

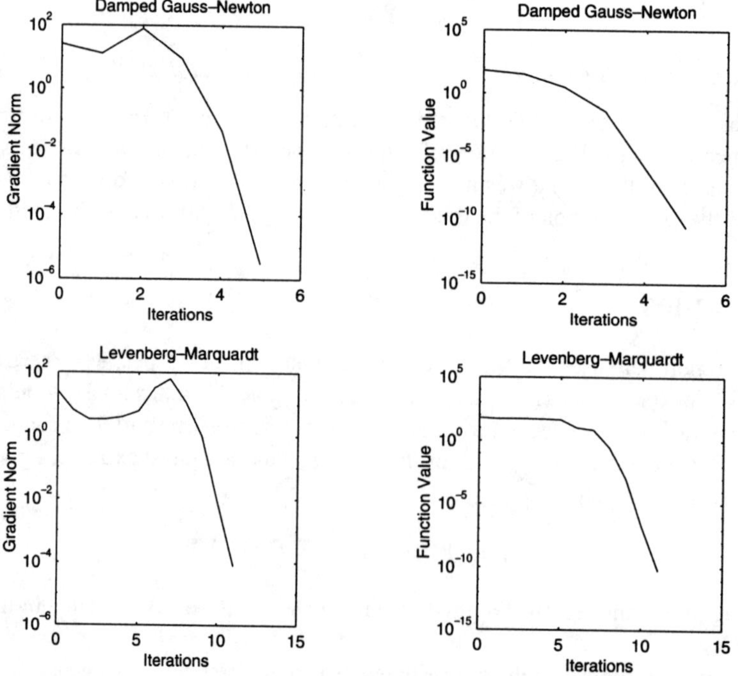

Figure 3.2: *Gauss–Newton and Levenberg–Marquardt for Parameter ID Problem*

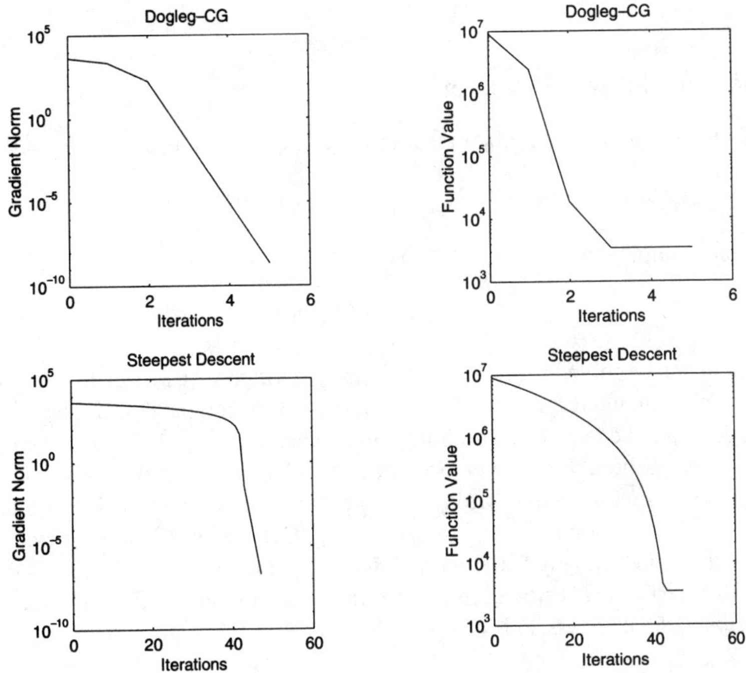

Figure 3.3: *Steepest Descent and Dogleg–CG for Discrete Control Problem*

3.4.1 Parameter Identification

We consider the problem from §2.6.1 except we use the initial data $x_0 = (5,5)^T$. Both the Gauss–Newton and Newton methods will fail to converge with this initial data without globalization (see Exercise 3.5.14). Newton's method has particular trouble with this problem because the Newton direction is not a descent direction in the early phases of the iteration. The termination criterion and difference increment for the finite difference Hessian was the same as for the computation in §2.6.1.

In Figure 3.1 we compare the performance of the Newton dogleg algorithm with the steepest descent algorithm. Our implementation of the classical dogleg in `ntrust` uses the standard values

(3.51) $$\omega_{down} = .5, \omega_{up} = 2, \mu_0 = \mu_{low} = .25, \text{ and } \mu_{high} = .75.$$

The plots clearly show the locally superlinear convergence of Newton's method and the linear convergence of steepest descent. However, the graphs do not completely show the difference in computational costs. In terms of gradient evaluations, steepest descent was marginally better than the Newton dogleg algorithm, requiring 50 gradients as opposed to 55 (which includes those needed for the 18 difference Hessian evaluations) for the Newton dogleg algorithm. However, the steepest descent algorithm required 224 function evaluations, while the Newton dogleg needed only 79. As a result, the Newton dogleg code was much more efficient, needing roughly 5 million floating point operations instead of the 10 million needed by the steepest descent code.

In Figure 3.2 we plot the performance of the damped Gauss–Newton and Levenberg–Marquardt algorithms. These exploit the least squares structure of the problem and are locally superlinearly convergent because this is a zero residual problem. They also show that algorithms that effectively exploit the structure of the least squares problem are much more efficient. Gauss–Newton required 6 gradient evaluations, 14 function evaluations, and 750 thousand floating point operations, and Levenberg–Marquardt required 12 gradients, 23 functions, and 1.3

million floating point operations.

3.4.2 Discrete Control Problem

We consider the discrete control problem from §1.6.1 with $N = 400$, $T = 1$, $y_0 = 0$,

$$L(y, u, t) = (y - 3)^2 + .5 * u^2, \text{ and } \phi(y, u, t) = uy + t^2.$$

We chose the poor initial iterate

$$u_0(t) = 5 + 300 \sin(20\pi t).$$

This problem can be solved very efficiently with Algorithm `cgtrust`. In our implementation we use the same parameters from (3.51). In Figure 3.3 we compare the dogleg–CG iteration with steepest descent. We terminated both iterations when $\|\nabla f\| < 10^{-8}$. For the dogleg–CG code we used $\eta = .01$ throughout the entire iteration and an initial trust region radius of $\|u_0\|$. The steepest descent computation required 48 gradient evaluations, 95 function evaluations, and roughly 1 million floating point operations, and dogleg–CG needed 17 gradient evaluations, 21 function evaluations, and roughly 530 thousand floating point operations. Note that the steepest descent algorithm performed very well in the terminal phase of the iteration. The reason for this is that, in this example, the Hessian is near the identity.

3.5 Exercises on Global Convergence

3.5.1. Let F be a nonlinear function from $R^N \to R^N$. Let

$$f(x) = \|F(x)\|^2/2.$$

What is ∇f? When is the Newton step for the nonlinear equation $F(x) = 0$,

$$d = -F'(x)^{-1}F(x),$$

a descent direction for f at x?

3.5.2. Prove Lemma 3.2.1.

3.5.3. Implement Algorithm `steep` without the scaling fixup in (3.50). Apply this crippled algorithm to the control problem example from §3.4.2. What happens and why?

3.5.4. Show that if f is a convex quadratic then f is bounded from below.

3.5.5. Verify (3.40).

3.5.6. Show that the Levenberg–Marquardt steps computed by (3.20) and (3.21) are the same.

3.5.7. Prove Theorem 3.2.7.

3.5.8. Complete the proof of Theorem 3.3.2.

3.5.9. Prove Theorem 3.3.4.

3.5.10. Look at the trust region algorithm for nonlinear equations from [218] or [84]. What are the costs of that algorithm that are not present in a line search? When might this trust region approach have advantages for solving nonlinear equations? Could it be implemented inexactly?

3.5.11. The double dogleg method [80], [84] puts a new node on the dogleg path in the Newton direction, thereby trying more aggressively for superlinear convergence. Implement this method, perhaps by modifying the MATLAB code `ntrust.m`, and compare the results with the examples in §3.4. Prove convergence results like Theorems 3.3.7 and 3.3.9 for this method.

3.5.12. In [51] a trust region algorithm was proposed that permitted inaccurate gradient computations, with the relative accuracy being tightened as the iteration progresses. Look at [51] and try to design a similar algorithm based on the line search paradigm. What problems do you encounter? How do you solve them?

3.5.13. Suppose one modifies Algorithm `trtest` by not resolving the trust region problem if the trial point is rejected, but instead performing a line search from x_t, and setting $\Delta = \|x_+ - x_c\|$, where x_+ is the accepted point from the line search. Discuss the merits of this modification and any potential problems. See [209] for the development of this idea.

3.5.14. Write programs for optimization that take full Gauss–Newton or Newton steps (you can cripple the MATLAB codes `gaussn.m` and `ntrust.m` for this). Apply these codes to the parameter identification problem from §3.4.1. What happens?

3.5.15. Write a nonlinear CG code and apply it to the problems in §3.4. Try at least two ways to manage the line search. How important are the (strong) Wolfe conditions?

3.5.16. Discuss the impact of using a difference Hessian in Algorithm `trcg`. How will the global convergence of Algorithm `cgtrust` be affected? How about the local convergence? Consider the accuracy in the evaluation of ∇f in your results.

3.5.17. Without looking at [247] describe in general terms how the proof of Theorem 3.3.1 should be modified to prove Theorem 3.3.10. Then examine the proof in [247] to see if you left anything out.

Chapter 4

The BFGS Method

Quasi-Newton methods update an approximation of $\nabla^2 f(x^*)$ as the iteration progresses. In general the transition from current approximations x_c and H_c of x^* and $\nabla^2 f(x^*)$ to new approximations x_+ and H_+ is given (using a line search paradigm) by the following steps:

1. Compute a search direction $d = -H_c^{-1}\nabla f(x_c)$.
2. Find $x_+ = x_c + \lambda d$ using a line search to insure sufficient decrease.
3. Use x_c, x_+, and H_c to *update* H_c and obtain H_+.

The way in which H_+ is computed determines the method.

The BFGS (Broyden, Fletcher, Goldfarb, Shanno) [36], [103], [124], [237] method, which is the focus of this chapter, and the other methods we will mention in §4.3 are also called *secant methods* because they satisfy the *secant equation*

$$H_+ s = y. \tag{4.1}$$

In (4.1)

$$s = x_+ - x_c \text{ and } y = \nabla f(x_+) - \nabla f(x_c).$$

If $N = 1$, all secant methods reduce to the classical *secant method* for the single nonlinear equation $f'(x) = 0$, i.e.,

$$x_+ = x_c - \frac{f'(x_c)(x_c - x_-)}{f'(x_c) - f'(x_-)}, \tag{4.2}$$

where x_- is the iterate previous to x_c.

The standard quasi-Newton update for nonlinear equations is Broyden's [34] method, a rank-one update,

$$H_+ = H_c + \frac{(y - H_c s)s^T}{s^T s}. \tag{4.3}$$

Broyden's method does not preserve the structural properties needed for line search methods in optimization, namely, symmetry and positive definiteness, and could, in fact, encourage convergence to a local maximum. For that reason quasi-Newton methods in optimization are more complex than those used for nonlinear equations. The methods of analysis and implementation are more complex as well.

In this chapter we will concentrate on the BFGS method [36], [103], [124], [237], which is the rank-two update

$$H_+ = H_c + \frac{yy^T}{y^T s} - \frac{(H_c s)(H_c s)^T}{s^T H_c s}. \tag{4.4}$$

We will briefly discuss other updates and variations that exploit problem structure in §4.3.

4.1 Analysis

This section begins with some simple observations on nonsingularity and positivity of the update.

It is very useful for both theory and practice to express (4.4) in terms of the inverse matrices. The formula we use in this book is Lemma 4.1.1.

LEMMA 4.1.1. *Let H_c be spd, $y^T s \neq 0$, and H_+ given by (4.4). Then H_+^{-1} is nonsingular and*

$$H_+^{-1} = \left(I - \frac{sy^T}{y^T s}\right) H_c^{-1} \left(I - \frac{ys^T}{y^T s}\right) + \frac{ss^T}{y^T s}. \tag{4.5}$$

Proof. See exercise 4.5.2. □

LEMMA 4.1.2. *Let H_c be spd, $y^T s > 0$, and H_+ given by (4.4). Then H_+ is spd.*

Proof. Positivity of H_c and $y^T s \neq 0$ imply that for all $z \neq 0$,

$$z^T H_+ z = \frac{(z^T y)^2}{y^T s} + z^T H_c z - \frac{(z^T H_c s)^2}{s^T H_c s}.$$

Using the symmetry and positivity of H_c, we have

$$(z^T H_c s)^2 \leq (s^T H_c s)(z^T H_c z),$$

with equality only if $z = 0$ or $s = 0$, and, therefore, since $z, s \neq 0$ and $y^T s > 0$,

$$z^T H_+ z > \frac{(z^T y)^2}{y^T s} \geq 0,$$

as asserted. □

If $y^T s \leq 0$ the update is considered a failure.

4.1.1 Local Theory

The local theory [37] requires accurate initial approximations to both x^* and $\nabla^2 f(x^*)$. The statement of the convergence result is easy to understand.

THEOREM 4.1.3. *Let the standard assumptions hold. Then there is δ such that if*

$$\|x_0 - x^*\| \leq \delta \text{ and } \|H_0 - \nabla^2 f(x^*)\| \leq \delta,$$

then the BFGS iterates are defined and converge q-superlinearly to x^*.

Technical Details

The proof of Theorem 4.1.3 is technical and we subdivide it into several lemmas. Our proof is a hybrid of ideas from [37], [135], and [154]. Similar to other treatments of this topic [45] we begin with the observation (see §2.5.2) that one may assume $\nabla^2 f(x^*) = I$ for the convergence analysis.

LEMMA 4.1.4. *Let the standard assumptions hold and let*

$$\hat{f}(y) = f(Ay),$$

where $A = (\nabla^2 f(x^*))^{-1/2}$. *Let* x_c *and* H_c *be given and let* $\hat{x}_c = A^{-1}x_c$ *and* $\hat{H}_c = AH_cA$. *Then the BFGS updates* (x_+, H_+) *for* f *and* $(\hat{x}_+, \hat{H}_+)$ *for* $\hat{f}$ *are related by*

$$\hat{x}_+ = A^{-1}x_+ \text{ and } \hat{H}_+ = AH_+A.$$

In particular, the BFGS sequence for f *exists (i.e.,* H_n *is spd for all* n*) if and only if the BFGS sequence for* $\hat{f}$ *does and the convergence of* $\{x_n\}$ *is q-superlinear if and only if the convergence of* $\{\hat{x}_n\}$ *is.*

Proof. The proof is a simple calculation and is left for exercise 4.5.3. □

Hence we can, with no loss of generality, assume that $\nabla^2 f(x^*) = I$, for if this is not so, we can replace f by $\hat{f}$ and obtain an equivalent problem for which it is.

Keeping in mind our assumption that $\nabla^2 f(x^*) = I$, we denote errors in the inverse Hessian by

$$E = H^{-1} - \nabla^2 f(x^*)^{-1} = H^{-1} - I.$$

These errors satisfy a simple recursion [37].

LEMMA 4.1.5. *Let the standard assumptions hold. Let* H_c *be spd and*

$$x_+ = x_c - H_c^{-1}\nabla f(x_c).$$

Then there is δ_0 *such that if*

$$0 < \|x_c - x^*\| \le \delta_0 \text{ and } \|E_c\| \le \delta_0,$$

then $y^T s > 0$. *Moreover, if* H_+ *is the BFGS update of* H_c *then*

$$E_+ = (I - ww^T)E_c(I - ww^T) + \Delta, \tag{4.6}$$

where $w = s/\|s\|$ *and for some* $K_\Delta > 0$

$$\|\Delta\| \le K_\Delta \|s\|. \tag{4.7}$$

Proof. Let δ_0 be small enough so that $\nabla f(x_c) \ne 0$ if $x_c \ne x^*$. Theorem 1.2.1 implies that

$$\nabla f(x_c) = \int_0^1 \nabla^2 f(x^* + te_c)e_c\, dt = e_c + \Delta_1 e_c,$$

where Δ_1 is the matrix given by

$$\Delta_1 = \int_0^1 (\nabla^2 f(x^* + te_c) - I)\, dt.$$

Clearly

$$\|\Delta_1\| \le \gamma\|e_c\|/2,$$

and

$$s = -H_c^{-1}\nabla f(x_c) = -(I + E_c)(I + \Delta_1)e_c.$$

Therefore,

$$\|e_c\|(1 - \delta_0)(1 - \gamma\delta_0/2) \le \|s\| \le \|e_c\|(1 + \delta_0)(1 + \gamma\delta_0/2)$$

and hence

$$0 < \|e_c\|/2 \le \|s\| \le 2\|e_c\| \tag{4.8}$$

if, say,

$$\delta_0 \le \min(1/4, 1/(2\gamma)). \tag{4.9}$$

We will assume that (4.9) holds for the rest of this section.

The standard assumptions, our assumption that $\nabla^2 f(x^*) = I$, and the fundamental theorem of calculus imply that

$$(4.10)\qquad \begin{aligned} y &= \nabla f(x_+) - \nabla f(x_c) = \int_0^1 \nabla^2 f(x_c + ts)s\,dt \\ &= s + \int_0^1 (\nabla^2 f(x_c + ts) - I)s\,dt = s + \Delta_2 s, \end{aligned}$$

where Δ_2 is the matrix given by

$$\Delta_2 = \int_0^1 (\nabla^2 f(x_c + ts) - I)\,dt.$$

The standard assumptions imply that $\|\Delta_2\| \le \gamma(\|e_+\| + \|e_c\|)/2$. Hence, (4.8) implies that

$$(4.11)\qquad y^T s = s^T s + (\Delta_2 s)^T s \ge \|s\|^2(1 - 3\gamma\|e_c\|/2)\|s\|^2(1 - 3\gamma\delta_0/2) > 0$$

provided $\delta_0 < 2\gamma/3$. We have that

$$(4.12)\qquad \frac{sy^T}{y^T s} = \frac{ss^T + s(\Delta_2 s)^T}{s^T s + (\Delta_2 s)^T s} = \frac{ss^T}{s^T s} - \Delta_3 = ww^T - \Delta_3,$$

where (see exercise 4.5.4), for some $C > 0$,

$$(4.13)\qquad \|\Delta_3\| \le C\|s\|.$$

Subtracting $(\nabla^2 f(x^*))^{-1} = I$ from (4.5) and using (4.12) gives us

$$\begin{aligned} E_+ &= (I - ww^T + \Delta_3)H_c^{-1}(I - ww^T + \Delta_3^T) + ww^T - I \\ &= (I - ww^T)(E_c + I)(I - ww^T) + ww^T - I + \Delta \\ &= (I - ww^T)E_c(I - ww^T) + \Delta, \end{aligned}$$

where

$$\Delta = \Delta_3 H_c^{-1}(I - ww^T + \Delta_3^T) + (I - ww^T)H_c^{-1}\Delta_3^T.$$

Therefore, if (4.9) holds then $1 + \delta_0 \le 3/2$ and

$$\begin{aligned} \|\Delta\| &\le (1 + \delta_0)\|\Delta_3\|(2 + \|\Delta_3\|) \le \|s\|3C(2 + C\|s\|)/2 \\ &\le 3C\|s\|(2 + 2C\delta_0)/2. \end{aligned}$$

Reduce δ_0 if necessary so that $2C\delta_0 \le 1$ and the proof is complete with $K_\Delta = 9C/2$. ☐

Lemma 4.1.5 implies that the approximate Hessians do not drift too far from the exact Hessian if the initial data are good. This property, called *bounded deterioration* in [37], will directly imply local q-linear convergence.

COROLLARY 4.1.6. *Let the assumptions of Lemma* 4.1.5 *hold and let* δ_0 *be as in the statement of Lemma* 4.1.5. *Then*

$$(4.14)\qquad \|E_+\| \le \|E_c\| + K_\Delta\|s\| \le \|E_c\| + K_\Delta(\|e_c\| + \|e_+\|).$$

Proof. The leftmost inequality follows from Lemma 4.1.5 and the fact that $I - ww^T$ is an orthogonal projection. The final inequality follows from the triangle inequality. □

We are now ready to prove local q-linear convergence. This is of interest in its own right and is a critical step in the superlinear convergence proof. Note that, unlike the statement and proof of Theorem 2.3.4, we do not express the estimates in terms of $\|H - \nabla^2 f(x^*)\| = \|H - I\|$ but in terms of $E = H^{-1} - I$. The two approaches are equivalent, since if $\|E\| \le \delta_\ell < 1/2$, then $\|H^{-1}\| < 3/2$ and the Banach lemma implies that $\|H\| \le 2$. Hence

$$\begin{aligned}\|H_n - I\|/2 &\le \|H_n\|^{-1}\|H_n - I\| \\ &\le \|H_n^{-1} - I\| = \|H_n^{-1}(H_n - I)\| \\ &\le \|H_n^{-1}\|\|H_n - I\| \le 3\|H_n - I\|/2.\end{aligned}$$

THEOREM 4.1.7. *Let the standard assumptions hold and let $\sigma \in (0,1)$. Then there is δ_ℓ such that if*

$$\|x_0 - x^*\| \le \delta_\ell \text{ and } \|H_0^{-1} - \nabla^2 f(x^*)^{-1}\| \le \delta_\ell, \tag{4.15}$$

then the BFGS iterates are defined and converge q-linearly to x^* with q-factor at most σ.

Proof. For $\hat{\delta}$ sufficiently small and

$$\|x_c - x^*\| \le \hat{\delta} \text{ and } \|E_c\| = \|H_c^{-1} - I\| \le \hat{\delta}, \tag{4.16}$$

the standard assumptions imply that there is $\bar{K}$ such that

$$\|e_+\| \le \bar{K}(\|E_c\|\|e_c\| + \|e_c\|^2)/2 \le \bar{K}\hat{\delta}\|e_c\|. \tag{4.17}$$

Reduce $\hat{\delta}$ if necessary so that $\bar{K}\hat{\delta} \le \sigma$ to obtain $\|e_+\| \le \sigma\|e_c\|$. The method of proof is to select δ_ℓ so that (4.16) is maintained for the entire iteration if the initial iterates satisfy (4.15).

With this in mind we set

$$\delta_\ell = \delta^*/2\left(1 + \frac{K_\Delta(1+\sigma)}{1-\sigma}\right)^{-1} < \delta^*/2 \tag{4.18}$$

where K_Δ is from Lemma 4.1.5. Now if $\|H_0 - I\| < \delta_\ell$ then

$$\|E_0\| \le \delta_\ell/(1-\delta_\ell) \le 2\delta_\ell < \delta^*$$

which is the estimate we need.

Now by Corollary 4.1.6

$$\|E_1\| \le \|E_0\| + K_\Delta(1+\sigma)\|e_0\|.$$

The proof will be complete if we can show that (4.15) and (4.18) imply that $\|E_n\| < \delta^*$ for all n. We do this inductively. If $\|E_n\| < \delta^*$ and $\|e_{j+1}\| \le \sigma\|e_j\|$ for all $j \le n$, then (4.14) implies that

$$\begin{aligned}\|E_{n+1}\| &\le \|E_n\| + K_\Delta(\|e_n\| + \|e_{n+1}\|) \le \|E_n\| + K_\Delta(1+\sigma)\|e_n\| \\ &\le \|E_n\| + K_\Delta(1+\sigma)\sigma^n\|e_0\| \le \|E_n\| + K_\Delta(1+\sigma)\sigma^n\delta_\ell \\ &\le \|E_0\| + \delta_\ell K_\Delta(1+\sigma)\sum_{j=0}^{n}\sigma^n \\ &= \delta_\ell\left(1 + \frac{K_\Delta(1+\sigma)}{1-\sigma}\right).\end{aligned}$$

We complete the induction and the proof by invoking (4.18) to conclude that $\|E_{n+1}\| \le \delta^*$. □

Proof of Theorem 4.1.3

The *Dennis–Moré condition* [82], [81] is a necessary and sufficient condition for superlinear convergence of quasi-Newton methods. In terms of the assumptions we make in this section, the condition is

$$\lim_{n\to\infty} \frac{\|E_n s_n\|}{\|s_n\|} = 0, \tag{4.19}$$

where $\{s_n\}$ is the sequence of steps and $\{E_n\}$ is the sequence of errors in the inverse Hessian. We will only state and prove the special case of the necessary condition that we need and refer the reader to [82], [81], [84], or [154] for more general proofs.

THEOREM 4.1.8. *Let the standard assumptions hold; let $\{H_n\}$ be a sequence of nonsingular $N \times N$ matrices satisfying*

$$\|H_n\| \le M \tag{4.20}$$

for some $M > 0$. Let $x_0 \in R^N$ be given and let $\{x_n\}_{n=1}^\infty$ be given by

$$x_{n+1} = x_n - H_n^{-1}\nabla f(x_n)$$

for some sequence of nonsingular matrices H_n. Then if $x_n \to x^*$ q-linearly, $x_n \ne x^*$ for any n, and (4.19) *holds then $x_n \to x^*$ q-superlinearly.*

Proof. We begin by invoking (4.10) to obtain

$$E_n s_n = (H_n^{-1} - I)s_n = (H_n^{-1} - I)(y_n - \Delta_2 s) = E_n y_n + O(\|s_n\|^2).$$

Convergence of x_n to x^* implies that $s_n \to 0$ and hence (4.19) can be written as

$$\lim_{n\to\infty} \frac{\|E_n y_n\|}{\|s_n\|} = 0, \tag{4.21}$$

where $y_n = \nabla f(x_{n+1}) - \nabla f(x_n)$.

Now let σ be the q-factor for the sequence $\{x_n\}$. Clearly

$$(1-\sigma)\|e_n\| \le \|s_n\| \le (1+\sigma)\|e_n\|.$$

Hence (4.21) is equivalent to

$$\lim_{n\to\infty} \frac{\|E_n y_n\|}{\|e_n\|} = 0. \tag{4.22}$$

Since $H_n^{-1}\nabla f(x_n) = -s_n$ and $s_n = y_n + O(\|s_n\|^2)$ we have

$$\begin{aligned} E_n y_n &= (H_n^{-1} - I)(\nabla f(x_{n+1}) - \nabla f(x_n)) \\ &= H_n^{-1}\nabla f(x_{n+1}) + s_n - y_n = H_n^{-1}\nabla f(x_{n+1}) + O(\|s_n\|^2) \\ &= H_n^{-1} e_{n+1} + O(\|e_n\|^2 + \|s_n\|^2) = H_n^{-1} e_{n+1} + O(\|e_n\|^2). \end{aligned}$$

Therefore, (4.22) implies that

$$\frac{\|E_n y_n\|}{\|e_n\|} = \frac{\|H_n^{-1} e_{n+1}\|}{\|e_n\|} + O(\|e_n\|) \ge M^{-1}\frac{\|e_{n+1}\|}{\|e_n\|} + O(\|e_n\|) \to 0$$

as $n \to \infty$, proving q-superlinear convergence. □

For the remainder of this section we assume that (4.15) holds and that δ_ℓ is small enough so that the conclusions of Theorem 4.1.7 hold for some $\sigma \in (0,1)$. An immediate consequence of this is that

$$\sum_{n=0}^{\infty} \|s_n\| < \infty. \tag{4.23}$$

The *Frobenius norm* of a matrix A is given by

$$\|A\|_F^2 = \sum_{i,j=1}^{N} (A)_{ij}^2. \tag{4.24}$$

It is easy to show that (see exercise 4.5.5) for any unit vector $v \in R^N$,

$$\|A(I - vv^T)\|_F^2 \leq \|A\|_F^2 - \|Av\|^2 \text{ and } \|(I - vv^T)A\|_F^2 \leq \|A\|_F^2. \tag{4.25}$$

We have, using (4.6), (4.7), and (4.25), that

$$\|E_{n+1}\|_F^2 \leq \|E_n\|_F^2 - \|E_n w_n\|^2 + O(\|s_n\|) = (1 - \theta_n^2)\|E_n\|_F^2 + O(\|s_n\|), \tag{4.26}$$

where $w_n = s_n / \|s_n\|$ and

$$\theta_n = \begin{cases} \dfrac{\|E_n w_n\|}{\|E_n\|_F} & \text{if } E_n \neq 0, \\ 1 & \text{if } E_n = 0. \end{cases}$$

Using (4.23) we see that for any $k \geq 0$,

$$\begin{aligned} \sum_{n=0}^{k} \theta_n^2 \|E_n\|_F^2 &\leq \sum_{n=0}^{k} \|E_n\|_F^2 - \|E_{n+1}\|_F^2 + O(1) \\ &= \|E_0\|_F^2 - \|E_{k+1}\|_F^2 + O(1) < \infty. \end{aligned}$$

Hence $\theta_n \|E_n\|_F \to 0$.

However,

$$\begin{aligned} \theta_n \|E_n\|_F &= \begin{cases} \|E_n w_n\| & \text{if } E_n \neq 0 \\ 0 & \text{if } E_n = 0 \end{cases} \\ &= \|E_n w_n\| = \frac{\|E_n s_n\|}{\|s_n\|}. \end{aligned}$$

Hence (4.19) holds. This completes the proof of Theorem 4.1.3.

4.1.2 Global Theory

If one uses the BFGS model Hessian in the context of Algorithm `optarm`, then Theorem 3.2.4 can be applied if the matrices $\{H_k\}$ remain bounded and well conditioned. However, even if a limit point of the iteration is a minimizer x^* that satisfies the standard assumptions, Theorem 3.2.4 does not guarantee that the iteration will converge to that point. The situation in which x is near x^* but H is not near $\nabla^2 f(x^*)$ is, from the point of view of the local theory, no better than that when x is far from x^*. In practice, however, convergence (often superlinear) is observed. The result in this section is a partial explanation of this.

Our description of the global theory, using the Armijo line search paradigm from Chapter 3, is based on [43]. We also refer the reader to [221], [45], and [269] for older results with a different line search approach. Results of this type require strong assumptions on f and the initial iterate x_0, but the reward is global and locally superlinear convergence for a BFGS–Armijo iteration.

ASSUMPTION 4.1.1. *The set*

$$D = \{x \mid f(x) \leq f(x_0)\}$$

is convex and f is Lipschitz twice continuously differentiable in D. Moreover, there are $\lambda_+ \geq \lambda_- > 0$ such that

$$\sigma(\nabla^2 f(x)) \subset [\lambda_-, \lambda_+]$$

for all $x \in D$.

Assumption 4.1.1 implies that f has a unique minimizer x^* in D and that the standard assumptions hold near x^*.

THEOREM 4.1.9. *Let Assumption 4.1.1 hold and let H_0 be spd. Then the BFGS–Armijo iteration converges q-superlinearly to x^*.*

The results for local and global convergence do not completely mesh. An implementation must allow for the fact that Assumption 4.1.1 may fail to hold, even near the root, and that $y^T s \leq 0$ is a possibility when far from the root.

4.2 Implementation

The two implementation issues that we must confront are storage of the data needed to maintain the updates and a strategy for dealing with the possibility that $y^T s \leq 0$. We address the storage question in §4.2.1. For the second issue, when $y^T s$ is not sufficiently positive, we restart the BFGS update with the identity. We present the details of this in §4.2.2. Our globalization approach, also given in §4.2.2, is the simplest possible, the Armijo rule as described in Chapter 3.

We choose to discuss the Armijo rule in the interest of simplicity of exposition. However, while the Armijo rule is robust and sufficient for most problems, more complex line search schemes have been reported to be more efficient [42], and one who seeks to write a general purpose optimization code should give careful thought to the best way to globalize a quasi-Newton method. In the case of BFGS, for example, one is always seeking to use a positive definite quadratic model, even in regions of negative curvature, and in such regions the approximate Hessian could be reinitialized to the identity more often than necessary.

4.2.1 Storage

For the present we assume that $y^T s > 0$. We will develop a storage-efficient way to compute the BFGS step using the history of the iteration rather than full matrix storage.

The implementation recommended here is one of many that stores the history of the iteration and uses that information recursively to compute the action of H_k^{-1} on a vector. This idea was suggested in [16], [186], [206], and other implementations may be found in [44] and [201]. All of these implementations store the iteration history in the pairs $\{s_k, y_k\}$ and we present a concrete example in Algorithm `bfgsrec`. A better, but somewhat less direct, way is based on the ideas in [91] and [275] and requires that only a single vector be stored for each iteration. We assume that we can compute the action of H_0^{-1} on a vector efficiently, say, by factoring H_0 at the outset of the iteration or by setting $H_0 = I$. We will use the BFGS formula from Lemma 4.1.1.

One way to maintain the update is to store the history of the iteration in the sequences of vectors $\{y_k\}$ and $\{s_k\}$ where

$$s_k = x_{k+1} - x_k \text{ and } y_k = \nabla f(x_{k+1}) - \nabla f(x_k).$$

If one has done this for $k = 0, \ldots, n-1$, one can compute the new search direction

$$d_n = -H_n^{-1} \nabla f(x_n)$$

by a recursive algorithm which applies (4.5).

Algorithm `bfgsrec` overwrites a given vector d with $H_n^{-1}d$. The storage needed is one vector for d and $2n$ vectors for the sequences $\{s_k, y_k\}_{k=0}^{n-1}$. A method for computing the product of H_0^{-1} and a vector must also be provided.

ALGORITHM 4.2.1. $\mathtt{bfgsrec}(n, \{s_k\}, \{y_k\}, H_0^{-1}, d)$

1. *If* $n = 0$, $d = H_0^{-1}d$; *return*
2. $\alpha = s_{n-1}^T d / y_{n-1}^T s$; $d = d - \alpha y_{n-1}$
3. *call* $\mathtt{bfgsrec}(n-1, \{s_k\}, \{y_k\}, H_0^{-1}, d)$
4. $d = d + (\alpha - (y_{n-1}^T d / y_{n-1}^T s_{n-1})) s_{n-1}$

Algorithm `bfgsrec` has the great advantage, at least in a language that efficiently supports recursion, of being very simple. More complex, but nonrecursive versions, have been described in [16], [201], and [44].

The storage cost of two vectors per iteration can be significant, and when available storage is exhausted one can simply discard the iteration history and restart with H_0. This approach, which we implement in the remaining algorithms in this section, takes advantage of the fact that if H_0 is spd then $-H_0^{-1}\nabla f(x)$ will be a descent direction, and hence useful for a line search. Another approach, called the *limited memory* BFGS [44], [207], [176], [201], keeps all but the oldest (s, y) pair and continues with the update. Neither of these approaches for control of storage, while essential in practice for large problems, has the superlinear convergence properties that the full-storage algorithm does.

At a cost of a modest amount of complexity in the formulation, we can reduce the storage cost to one vector for each iteration. The method for doing this in [275] begins with an expansion of (4.5) as

$$H_+^{-1} = H_c^{-1} + \alpha_0 s_c s_c^T + \beta_0((H_c^{-1}y_c)s_c^T + s_c(H_c^{-1}y_c)^T),$$

where

$$\alpha_0 = \frac{y_c^T s_c + y_c^T H_c^{-1} y_c}{(y_c^T s_c)^2} \text{ and } \beta_0 = \frac{-1}{y_c^T s_c}.$$

Now note that

$$H_c^{-1}y_c = H_c^{-1}\nabla f(x_+) - H_c^{-1}\nabla f(x_c) = H_c^{-1}\nabla f(x_+) + s_c/\lambda_c$$

and obtain

$$H_+^{-1} = H_c^{-1} + \alpha_1 s_c s_c^T + \beta_0(s_c(H_c^{-1}\nabla f(x_+))^T + (H_c^{-1}\nabla f(x_+))s_c^T), \tag{4.27}$$

where

$$\alpha_1 = \alpha_0 + 2\beta_0/\lambda_c.$$

Also

$$\begin{aligned} d_+ &= -H_+^{-1}\nabla f(x_+) \\ &= -\left(I - \frac{s_c y_c^T}{y_c^T s_c}\right) H_c^{-1} \left(I - \frac{y_c s_c^T}{y_c^T s_c}\right) \nabla f(x_+) - \frac{s_c s_c^T \nabla f(x_+)}{y_c^T s_c} \qquad (4.28) \\ &= A_c s_c + B_c H_c^{-1} \nabla f(x_+), \end{aligned}$$

where

$$A_c = \frac{y_c^T}{y_c^T s_c} H_c^{-1} \left(I - \frac{y_c s_c^T}{y_c^T s_c}\right) \nabla f(x_+) + \frac{s_c^T \nabla f(x_+)}{\lambda_c y_c^T s_c} \tag{4.29}$$

and

$$B_c = -1 + \frac{s_c^T \nabla f(x_+)}{y_c^T s_c}. \tag{4.30}$$

At this point we can compute d_+, and therefore λ_+ and s_+ using only $H_c^{-1}\nabla f(x_c)$. We do not need H_+ at all. We can now form H_+ with the new data for the next iterate and will show that we do not need to store the vectors $\{y_k\}$.

Since (verify this!) $B_c \neq 0$, we have

$$H_c^{-1}\nabla f(x_+) = -\frac{s_+}{B_c \lambda_+} + \frac{A_c s_c}{B_c}.$$

Combining this with (4.27) gives

$$H_+^{-1} = H_c^{-1} + \alpha_c s_c s_c^T + \beta_c (s_c s_+^T + s_+ s_c^T), \tag{4.31}$$

where

$$\alpha_c = \alpha_1 + 2\beta_0 A_c / B_c \text{ and } \beta_c = -\frac{\beta_0}{B_c \lambda_+}. \tag{4.32}$$

This leads to the expansion

$$H_{n+1}^{-1} = H_0^{-1} + \sum_{k=0}^{n} \alpha_k s_k s_k^T + \beta_k (s_k s_{k+1}^T + s_{k+1} s_k^T). \tag{4.33}$$

Upon reflection the reader will see that this is a complete algorithm. We can use (4.28) and H_n to compute d_{n+1}. Then we can compute λ_{n+1} and s_{n+1} and use them and (4.32) to compute α_n and β_n. This new data can be used to form H_{n+1}^{-1} with (4.33), which we can use to compute d_{n+2} and continue the iteration.

In this way only the steps $\{s_k\}$ and the expansion coefficients $\{\alpha_k\}$ and $\{\beta_k\}$ need be stored. Algorithm `bfgsopt` is an implementation of these ideas.

ALGORITHM 4.2.2. `bfgsopt`(x, f, ϵ)

1. $g = -\nabla f(x), n = 0.$
2. *While* $\|g\| > \epsilon$
   (a) *If* $n = 0$, $d_n = -H_0^{-1} g$
   otherwise compute A, B, and d_n *using* (4.28), (4.29), *and* (4.30).
   (b) *Compute* λ_n, s_n, *and* $x = x_{n+1}$ *with the Armijo rule.*
   (c) *If* $n > 0$ *compute* α_{n-1} *and* β_{n-1} *using* (4.32).
   (d) $g = -\nabla f(x), n = n + 1.$

4.2.2 A BFGS–Armijo Algorithm

In this section we present a simple implementation that shows how the theoretical results can be applied in algorithm design. Let H_+^{BFGS} be the BFGS update from H_c and define the two modified BFGS (MBFGS) updates by

$$H_+ = \begin{cases} H_+^{BFGS} & \text{if } y^T s > 0, \\ I & \text{if } y^T s \leq 0, \end{cases} \tag{4.34}$$

and

$$H_+ = \begin{cases} H_+^{BFGS} & \text{if } y^T s > 0, \\ H_c & \text{if } y^T s \le 0. \end{cases} \tag{4.35}$$

In the MBFGS1 method, (4.34), the model Hessian is reinitialized to I if $y^T s \le 0$. In the early phase of this iteration, where $\nabla^2 f$ may have negative eigenvalues, $y^T s \le 0$ is certainly possible and the search direction could be the steepest descent direction for several iterations.

An MBFGS2 step (4.35) keeps the history of the iteration even if $y^T s \le 0$. One view is that this approach keeps as much information as possible. Another is that once $y^T s \le 0$, the iteration history is suspect and should be thrown away. Both forms are used in practice. Our MATLAB code `bfgswopt` uses MFBGS1 and maintains an approximation to H^{-1} using Algorithm `bfgsopt`. We also guard against poor scaling by using (3.50).

4.3 Other Quasi-Newton Methods

The DFP (Davidon, Fletcher, Powell) update [71], [72], [105]

$$H_+ = H_c + \frac{(y - H_c s)y^T + y(y - H_c s)^T}{y^T s} - \frac{[(y - H_c s)^T y] y y^T}{(y^T s)^2} \tag{4.36}$$

has similar local convergence properties to BFGS but does not perform as well in practice [224], [225].

Two updates that preserve symmetry, but not definiteness, are the PSB (Powell symmetric Broyden) update [219],

$$H_+ = H_c + \frac{(y - H_c s)s^T + s(y - H_c s)^T}{s^T s} - \frac{[s^T (y - H_c s)] s s^T}{(s^T s)^2}, \tag{4.37}$$

and the symmetric rank-one (SR1) [35] update,

$$H_+ = H_c + \frac{(y - H_c s)(y - H_c s)^T}{(y - H_c s)^T s}. \tag{4.38}$$

By preserving the symmetry of the approximate Hessians, but not the positive definiteness, these updates present a problem for a line search globalization but an opportunity for a trust region approach. The SR1 update has been reported to outperform BFGS algorithms in certain cases [165], [41], [64], [65], [163], [258], [118], [250], [119], [268], [164], in which either the approximate Hessians can be expected to be positive definite or a trust region framework is used [41], [64], [65].

One may update the inverse of the SR1 approximate Hessian using the *Sherman–Morrison formula*, (4.39), a simple relation between the inverse of a nonsingular matrix and that of a rank-one update of that matrix [93], [239], [240], [14].

PROPOSITION 4.3.1. *Let H be a nonsingular $N \times N$ matrix and let $u, v \in R^N$. Then $H + uv^T$ is invertible if and only if $1 + v^T H^{-1} u \ne 0$. In this case*

$$(H + uv^T)^{-1} = \left(I - \frac{(H^{-1}u)v^T}{1 + v^T H^{-1} u} \right) H^{-1}. \tag{4.39}$$

The proof is simply a direct verification of (4.39).

The SR1 algorithm terminates in finitely many iterations for convex quadratic optimization problems [101]. Since the denominator $(y - H_c s)^T s$ could vanish, the update could completely fail and implementations must examine the denominator and take appropriate action if it is too small. This update does not enforce or require positivity of the approximate Hessian and has been used effectively to exploit negative curvature in a trust region context [165], [41].

For overdetermined nonlinear least squares problems one can try to approximate the second-order term in $\nabla^2 f$ while computing $R'^T R'$ exactly. Suppose

$$\nabla^2 f(x) \approx H = C(x) + A,$$

where the idea is that C, the computed part, is significantly easier to compute than A, the approximated part. This is certainly the case for nonlinear least squares, where $C = R'^T R'$. A quasi-Newton method that intends to exploit this structure will update A only; hence

$$H_+ = C(x_+) + A_+.$$

Superlinear convergence proofs require, in one way or another, that $H_+ s = y$. Therefore, in terms of A, one might require the update to satisfy

$$A_+ s = y^\# = y - C(x_+)s. \tag{4.40}$$

The definition of $y^\#$ given in (4.40) is called the *default choice* in [87]. This is not the only choice for $y^\#$, and one can prove superlinear convergence for this and many other choices [87], [84]. This idea, using several different updates, has been used in other contexts, such as optimal control [159], [164].

An algorithm of this type, using SR1 to update A and a different choice for $y^\#$, was suggested in [20] and [21]. The nonlinear least squares update from [77], [78], and [84] uses a DFP update and yet another $y^\#$ to compute A_+,

$$A_+ = A_c + \frac{(y^\# - A_c s){y^\#}^T + y^\#(y^\# - A_c s)^T}{{y^\#}^T s} - \frac{[(y^\# - A_c s)^T y^\#] y^\# {y^\#}^T}{({y^\#}^T s)^2}. \tag{4.41}$$

The application of this idea to large-residual least squares problems is not trivial, and scaling issues must be considered in a successful implementation.

Our proof of superlinear convergence can be applied to updates like (4.41). We state a special case of a result from [87] for the BFGS formulation

$$A_+ = A_c + \frac{y^\# {y^\#}^T}{{y^\#}^T s} - \frac{(A_c s)(A_c s)^T}{s^T A_c s}. \tag{4.42}$$

THEOREM 4.3.2. *Let the standard assumptions hold and assume that*

$$A^* = \nabla^2 f(x^*) - C(x^*)$$

is spd. Then there is δ such that if

$$\|x_0 - x^*\| \le \delta \text{ and } \|A_0 - A^*\| \le \delta,$$

then the quasi-Newton iterates defined by (4.42) *exist and converge q-superlinearly to x^*.*

This result can be readily proved using the methods in this chapter (see [159]).

Quasi-Newton methods can also be designed to take into account special structure, such as the sparsity pattern of the Hessian. One can update only those elements that are nonzero in the initial approximation to the Hessian, requiring that the secant equation $Hs = y$ holds. Such updates have been proposed and analyzed in varying levels of generality in [83], [87], [185], [238], [256], and [255].

Another approach is to use the dependency of f on subsets of the variables, a structure that is often present in discretizations of infinite-dimensional problems where coefficients of operators can be updated rather than entire matrix representations of those operators. We refer the reader to [133], [131], and [132] for an algebraic viewpoint based on finite-dimensional analysis and to [159], [157], [164], [160], and [163] for an operator theoretic description of these methods.

When applied to discretizations of infinite-dimensional optimization problems, quasi-Newton methods perform best when they also work well on the infinite-dimensional problem itself. Work on BFGS in Hilbert space can be found, for example, in [135], [158], and [159].

Quasi-Newton methods have been designed for underdetermined problems [184], and Broyden's method itself has been applied to linear least squares problems [111], [148].

4.4 Examples

The computations in this section were done with the MATLAB code `bfgswopt`. For the small parameter ID problem, where evaluation of f is far more expensive than the cost of maintaining or factoring the (very small!) approximate Hessian, one could also use a brute force approach in which H is updated and factored anew with each iteration.

4.4.1 Parameter ID Problem

We solve the parameter ID problem with the same data as in §3.4.1 using $H_0 = I$ as the initial Hessian. We compare the BFGS solution with the Gauss–Newton iteration from §3.4.1. From Figure 4.1 one can see the local superlinear convergence and the good performance of the line search. However, as one should expect, the Gauss–Newton iteration, being designed for small residual least squares problems, was more efficient. The Gauss–Newton iteration required 14 function evaluations, 6 gradients, and roughly 1.3 million floating point operations, while the BFGS–Armijo iteration needed 29 function evaluations, 15 gradients, and 3.8 million floating point operations.

4.4.2 Discrete Control Problem

We return to the example from §3.4.2. For our first example we use the initial iterate

$$u_0(t) = 10.$$

BFGS also requires an initial approximation to the Hessian and we consider two such approximations:

$$H_p = .25I \text{ and } H_g = I. \tag{4.43}$$

The Hessian for the continuous problem is a compact perturbation of the identity and the theory from [158] and [135] indicates that H_g is a much better approximate Hessian than H_p. The results in Figure 4.2 support that idea. For the better Hessian, one can see the concavity of superlinear convergence in the plot of the gradient norm. The computation for the better Hessian required 12 iterations and roughly 572 thousand floating point operations, while the one with the poor Hessian took 16 iterations and roughly 880 thousand floating point operations. Stepsize reductions were not required for the good Hessian and were needed four times during the iteration for the poor Hessian. However, the guard against poor scaling (3.50) was needed in both cases.

When we used the same poor initial iterate that we used in §3.4.2

$$u_0(t) = 5 + 300\sin(20\pi t)$$

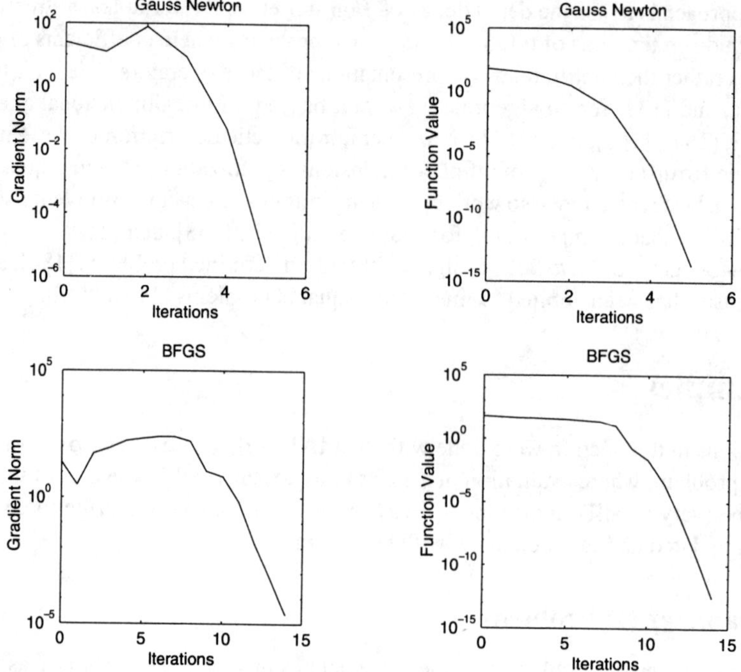

Figure 4.1: *BFGS–Armijo and Gauss–Newton for the Parameter ID Problem*

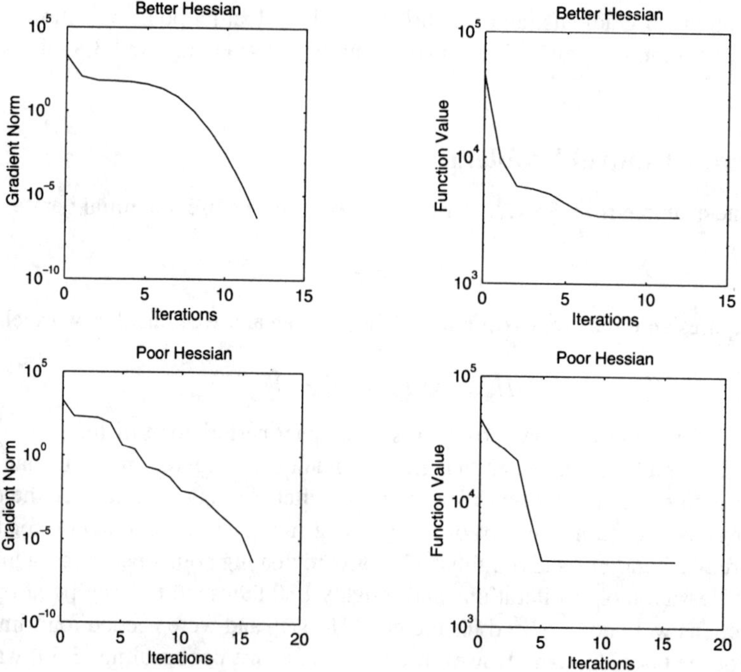

Figure 4.2: *BFGS–Armijo for Discrete Control Problem*

and allocated 50 vectors to Algorithm `bfgsopt`, there was no longer a benefit to using the good Hessian. In fact, as is clear from Figure 4.3 the poor Hessian produced a more rapidly convergent iteration.

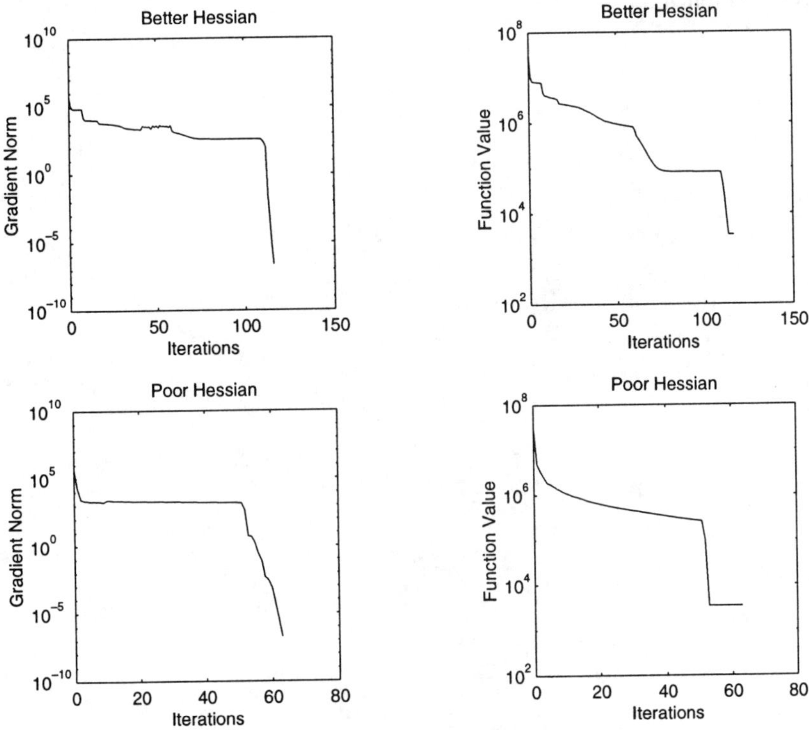

Figure 4.3: *BFGS–Armijo for Discrete Control Problem: Poor Initial Iterate*

4.5 Exercises on BFGS

4.5.1. Use the secant method (4.2) with initial data $x_{-1} = 1$ and $x_0 = .9$ to minimize $f(x) = x^4$. Explain the convergence of the iteration.

4.5.2. Prove Lemma 4.1.1. It might help to use the secant equation.

4.5.3. Prove Lemma 4.1.4.

4.5.4. Verify (4.13) and compute the constant C.

4.5.5. Prove (4.25).

4.5.6. As an exercise in character building, implement Algorithm `bfgsrec` nonrecursively.

4.5.7. Show how the Sherman–Morrison formula can be used to implement the SR1 update in such a way that only one vector need be stored for each iterate.

4.5.8. State and prove a local convergence theorem for DFP and/or PSB.

4.5.9. Implement the DFP and PSB update and compare their performance with BFGS on the examples from §4.4.

4.5.10. Show that, for positive definite quadratic problems, the BFGS method with an *exact line search* (i.e., one that finds the minimum of f in the search direction) is the same as CG [201], [200].

4.5.11. Prove Theorem 4.3.2.

Chapter 5

Simple Bound Constraints

5.1 Problem Statement

The goal of this chapter is to show how the techniques of Chapters 2, 3, and 4 can be used to solve a simple constrained optimization problem. The algorithm we suggest at the end in §5.5.3 is a useful extension of the BFGS–Armijo algorithm from Chapter 4. We will continue this line of development when we solve noisy problems in Chapter 7.

Let $\{L_i\}_{i=1}^N$ and $\{U_i\}_{i=1}^N$ be sequences of real numbers such that

$$-\infty < L_i < U_i < +\infty. \tag{5.1}$$

The bound constrained optimization problem is to find a local minimizer x^* of a function f of N variables subject to the *constraint* that

$$x^* \in \Omega = \{x \in R^N \mid L_i \leq (x)_i \leq U_i\}. \tag{5.2}$$

By this we mean that x^* satisfies

$$f(x^*) \leq f(x) \text{ for all } x \in \Omega \text{ near } x^*. \tag{5.3}$$

It is standard to express this problem as

$$\min_{x \in \Omega} f(x) \tag{5.4}$$

or as $\min_\Omega f$. The set Ω is called the *feasible set* and a point in Ω is called a *feasible point.*

Because the set Ω is compact there is always a solution to our minimization problem [229].

The inequalities $L_i \leq (x)_i \leq U_i$ are called *inequality constraints* or simply *constraints.* We will say that the ith constraint is *active* at $x \in \Omega$ if either $(x)_i = L_i$ or $(x)_i = U_i$. If the ith constraint is not active we will say that it is *inactive*. The set of indices i such that the ith constraint is active (inactive) will be called the set of *active (inactive) indices* at x.

We will write $\mathcal{A}(x)$ and $\mathcal{I}(x)$ for the active and inactive sets at x.

5.2 Necessary Conditions for Optimality

For a continuously differentiable function of one variable, the necessary conditions for unconstrained optimality at x^* are simply $f'(x^*) = 0$ and, if f is twice continuously differentiable, $f''(x^*) \geq 0$. A bound constrained problem in one variable restricts the domain of f to an interval $[a, b]$, and the necessary conditions must be changed to admit the possibility that the minimizer is one of the endpoints. If $x^* = a$ is a local minimizer, then it need not be the case that $f'(a) = 0$; however, because a is a local minimizer, $f(x) \geq f(a)$ for all $a \leq x$ sufficiently near a. Hence $f'(a) \geq 0$. Nothing, however, can be said about f''. Similarly, if $x^* = b$ is a local minimizer, $f'(b) \leq 0$. If f is differentiable on $[a, b]$ (i.e., on an open set containing $[a, b]$), then the necessary conditions for all three possibilities, $x^* = a$, $x^* = b$, and $a < x^* < b$ can be neatly expressed by the following theorem.

THEOREM 5.2.1. *Let f be a continuously differentiable function of one variable on the interval $[a, b]$. Let x^* be a local minimum of f on $[a, b]$. Then*

$$f'(x^*)(x - x^*) \geq 0 \text{ for all } x \in [a, b] \tag{5.5}$$

and, if f is twice continuously differentiable on $[a, b]$,

$$f''(x^*)(x^* - a)(b - x^*) \geq 0. \tag{5.6}$$

The analogue (5.5) is expressed by the idea of *stationarity*.

DEFINITION 5.2.1. *A point $x^* \in \Omega$ is* stationary *for problem* (5.4) *if*

$$\nabla f(x^*)^T(x - x^*) \geq 0 \text{ for all } x \in \Omega. \tag{5.7}$$

As in the unconstrained case, stationary points are said to satisfy the *first-order necessary conditions*.

The fact that optimality implies stationarity is proved with Taylor's theorem just as it was in the unconstrained case.

THEOREM 5.2.2. *Let f be continuously differentiable on Ω and let x^* be a solution of problem* (5.4). *Then x^* is a stationary point for problem* (5.4).

Proof. Let x^* be a solution of problem (5.4) and let $y \in \Omega$. As Ω is convex, the line segment joining x^* and y is entirely in Ω. Hence, the function

$$\phi(t) = f(x^* + t(y - x^*))$$

is defined for $t \in [0, 1]$ and has a local minimum at $t = 0$. Therefore, by Theorem 5.2.1

$$0 \leq \phi'(0) = \nabla f(x^*)^T(y - x^*)$$

as asserted. □

The case of a function of a single variable is less useful in explaining the role of the second derivative. However, we can get a complete picture by looking at functions of two variables. To illustrate the ideas we let $N = 2$ and let f be a twice Lipschitz continuously differentiable function on $\Omega = [0, 1] \times [0, 1]$. If x^* is a solution of (5.4) and no constraints are active, then $\nabla^2 f(x^*)$ is positive semidefinite by the same arguments used in the unconstrained case. If one or more constraints are active, however, then, just as in the one variable case, one cannot draw conclusions about the positivity of $\nabla^2 f(x^*)$. Suppose the minimizer is at $x^* = (\xi, 0)$ with $0 < \xi < 1$. While nothing can be said about $\partial^2 f(x^*)/\partial x_2^2$, the function $\phi(t) = f(t, 0)$, defined on $[0, 1]$, must satisfy

$$\phi''(\xi) = \partial^2 f(x^*)/\partial x_1^2 \geq 0.$$

Hence, second partials in directions corresponding to the inactive constraints must be nonnegative, while nothing can be said about directions corresponding to active constraints.

We define the *reduced Hessian* to help make this idea precise.

DEFINITION 5.2.2. *Let f be twice differentiable at $x \in \Omega$. The reduced Hessian $\nabla^2_R f(x)$ is the matrix whose entries are*

$$(\nabla^2_R f(x))_{ij} = \begin{cases} \delta_{ij} & \text{if } i \in \mathcal{A}(x) \text{ or } j \in \mathcal{A}(x), \\ (\nabla^2 f(x))_{ij} & \text{otherwise.} \end{cases} \tag{5.8}$$

We can now present the second-order necessary conditions.

THEOREM 5.2.3. *Let f be twice Lipschitz continuously differentiable and let x^* be the solution of problem* (5.4). *Then the reduced Hessian $\nabla^2_R f(x^*)$ is positive semidefinite.*

Proof. Assume that there are M inactive indices and $N - M$ active indices. We partition $x \in \Omega$, reordering the variables if needed, into $x = (\xi, \zeta)$ with ξ corresponding to the inactive indices and ζ to the active. The map

$$\phi(\xi) = f(\xi, \zeta^*)$$

has an unconstrained local minimizer at $\xi^* \in R^M$ and hence $\nabla^2 \phi$ is positive semidefinite. Since the reduced Hessian can be written as

$$\nabla^2_R f(x^*) = \begin{pmatrix} \nabla^2 \phi(x^*) & 0 \\ 0 & I \end{pmatrix}$$

if the variables are partitioned in this way, the proof is complete. □

We let $\mathcal{P}$ denote the *projection* onto Ω, that is, the map that takes x into the nearest point (in the l^2-norm) in Ω to x. We have that

$$\mathcal{P}(x)_i = \begin{cases} L_i & \text{if } (x)_i \le L_i, \\ (x)_i & \text{if } L_i < (x)_i < U_i, \\ U_i & \text{if } (x)_i \ge U_i. \end{cases} \tag{5.9}$$

Theorem 5.2.4 states our final necessary condition; we defer the proof to §5.4.4.

THEOREM 5.2.4. *Let f be continuously differentiable. A point $x^* \in \Omega$ is stationary for problem* (5.4) *if and only if*

$$x^* = \mathcal{P}(x^* - \lambda \nabla f(x^*)) \tag{5.10}$$

for all $\lambda \ge 0$.

5.3 Sufficient Conditions

With the definition of the reduced Hessian in hand, the sufficient conditions are easy to formulate. We begin by strengthening the notion of stationarity. If x^* is stationary, $i \in \mathcal{I}(x^*)$, and e_i is a unit vector in the ith coordinate direction, then $x^* \pm te_i \in \Omega$ for all t sufficiently small. Since

$$\frac{df(x^* \pm te_i)}{dt} = \pm \nabla f(x^*)^T e_i \ge 0,$$

therefore

$$(\nabla f(x^*))_i = 0 \text{ for all } i \in \mathcal{I}(x^*).$$

We will use the concept of nondegenerate stationary point or strict complementarity in our formulation of the sufficient conditions.

DEFINITION 5.3.1. *A point $x^* \in \Omega$ is a* nondegenerate stationary point *for problem* (5.4) *if x^* is a stationary point and*

$$(\nabla f(x^*))_i \neq 0 \text{ for all } i \in \mathcal{A}(x^*). \tag{5.11}$$

If x^* is also a solution of problem (5.4) *we say that x^* is a* nondegenerate local minimizer.

Our nondegeneracy condition is also referred to as *strict complementarity*.

If $\mathcal{S}$ is any set of indices define

$$(\mathcal{P}_{\mathcal{S}}x)_i = \begin{cases} (x)_i, & i \in \mathcal{S}, \\ 0, & i \notin \mathcal{S}. \end{cases}$$

Nondegeneracy is important not only in the formulation of sufficient conditions but also in the design of termination criteria. The first step in the use of nondegeneracy is Lemma 5.3.1.

LEMMA 5.3.1. *Let x^* be a nondegenerate stationary point. Assume that $\mathcal{A} = \mathcal{A}(x^*)$ is not empty. Then there is σ such that*

$$\nabla f(x^*)^T(x - x^*) = \nabla f(x^*)^T \mathcal{P}_{\mathcal{A}}(x - x^*) \geq \sigma \|\mathcal{P}_{\mathcal{A}}(x - x^*)\|$$

for all $x \in \Omega$.

Proof. If $i \in \mathcal{A}$ then nondegeneracy and stationarity imply that there is $\sigma > 0$ such that either

$$(x_i^*) = L_i \text{ and } (\nabla f(x^*))_i \geq \sigma \text{ or } (x_i^*) = U_i \text{ and } (\nabla f(x^*))_i \leq -\sigma.$$

If $x \in \Omega$ then for all $i \in \mathcal{A}$,

$$(\nabla f(x^*))_i(x - x^*)_i \geq \sigma|(x - x^*)_i|.$$

Therefore, since $\|x\|_1 \geq \|x\|_2$,

$$\nabla f(x^*)^T \mathcal{P}_{\mathcal{A}}(x - x^*) \geq \sigma \|\mathcal{P}_{\mathcal{A}}(x - x^*)\|,$$

as asserted. □

For a nondegenerate stationary point the sufficiency conditions are very similar to the unconstrained case.

THEOREM 5.3.2. *Let $x^* \in \Omega$ be a nondegenerate stationary point for problem* (5.4). *Let f be twice differentiable in a neighborhood of x^* and assume that the reduced Hessian at x^* is positive definite. Then x^* is a solution of problem* (5.4) *(and hence a nondegenerate local minimizer).*

Proof. Let $x \in \Omega$ and define $\phi(t) = f(x^* + t(x - x^*))$. We complete the proof by showing that either (i) $\phi'(0) > 0$ or (ii) $\phi'(0) = 0, \phi''(0) > 0$. Let $e = x - x^*$ and note that

$$\phi'(0) = \nabla f(x^*)^T e = \nabla f(x^*)^T(\mathcal{P}_{\mathcal{A}}e + \mathcal{P}_{\mathcal{I}}e).$$

Stationarity implies that $\nabla f(x^*)^T \mathcal{P}_{\mathcal{I}}e = 0$. If $\mathcal{P}_{\mathcal{A}}e \neq 0$ then nondegeneracy implies that

$$\nabla f(x^*)^T \mathcal{P}_{\mathcal{A}}e > 0$$

and hence (i) holds. If $\mathcal{P}_{\mathcal{A}}e = 0$ then

$$\phi''(0) = (x - x^*)^T \mathcal{P}_{\mathcal{I}} \nabla^2 f(x^*) \mathcal{P}_{\mathcal{I}}(x - x^*) = (x - x^*)^T \nabla_R^2 f(x^*)(x - x^*) > 0,$$

proving (ii). □

5.4 The Gradient Projection Algorithm

The gradient projection algorithm is the natural extension of the steepest descent algorithm to bound constrained problems. It shares all the advantages and disadvantages of that algorithm. Our approach follows that of [18]. Given a current iterate x_c the new iterate is

$$x_+ = \mathcal{P}(x_c - \lambda \nabla f(x_c)),$$

where λ is a steplength parameter given by the Armijo rule or some other line search scheme. In this section we will restrict our attention to the simplest form of the Armijo rule. In order to implement any line search scheme, we must specify what we mean by sufficient decrease. For $\lambda > 0$ define

$$x(\lambda) = \mathcal{P}(x - \lambda \nabla f(x)). \tag{5.12}$$

For bound constrained problems we will express the *sufficient decrease* condition for line searches (compare with (3.4)) as

$$f(x(\lambda)) - f(x) \leq \frac{-\alpha}{\lambda} \|x - x(\lambda)\|^2. \tag{5.13}$$

As with (3.4), α is a parameter and is typically set to 10^{-4} [84].

The general algorithmic description follows in Algorithm 5.4.1.

ALGORITHM 5.4.1. gradproj$(x, f, nmax)$

1. *For* $n = 1, \ldots, nmax$
   (a) *Compute* f *and* ∇f*; test for termination.*
   (b) *Find the least integer* m *such that* (5.13) *holds for* $\lambda = \beta^m$.
   (c) $x = x(\lambda)$.
2. *If* $n = nmax$ *and the termination test is failed, signal failure.*

The next step is to elaborate on the termination criterion.

5.4.1 Termination of the Iteration

The termination criterion for unconstrained optimization that we have used previously must be modified if we are to properly take the constraints into account. ∇f need not be zero at the solution, but a natural substitute is to terminate the iteration if the difference between x and $x(1)$ is small. As in the case of unconstrained optimization or nonlinear equations, we must invoke the sufficient conditions to show that such a termination criterion will accurately measure the error.

As usual, we let $e = x - x^*$.

We begin with a lemma that connects the active and inactive sets at a nondegenerate local minimizer with nearby points.

LEMMA 5.4.1. *Let* f *be twice continuously differentiable on* Ω *and let* x^* *be a nondegenerate stationary point for problem* (5.4). *Let* $\lambda \in (0, 1]$. *Then for* x *sufficiently near* x^*,

1. $\mathcal{A}(x) \subset \mathcal{A}(x^*)$ *and* $(x)_i = (x^*)_i$ *for all* $i \in \mathcal{A}(x)$.
2. $\mathcal{A}(x(\lambda)) = \mathcal{A}(x^*)$ *and* $(x(\lambda))_i = (x^*)_i$ *for all* $i \in \mathcal{A}(x^*)$.

Proof. Let

$$\mathcal{A}^* = \mathcal{A}(x^*), \mathcal{I}^* = \mathcal{I}(x^*), \mathcal{A} = \mathcal{A}(x), \text{ and } \mathcal{I} = \mathcal{I}(x).$$

Let

$$\delta_1 = \min_{i \in \mathcal{I}^*}\{(U_i - (x^*)_i), ((x^*)_i - L_i), (U_i - L_i)/2\}.$$

If $i \in \mathcal{I}^*$ and $\|e\| < \delta_1$ then $L_i < (x)_i < U_i$. Hence,

$$\mathcal{I}^* \subset \mathcal{I}$$

proving the first assertion that $\mathcal{A} \subset \mathcal{A}^*$. Moreover, since

$$\|e\| < \delta_1 \le \min\{(U_i - L_i)/2\},$$

then $(x)_i = (x^*)_i$ for all $i \in \mathcal{A}$.

Now let $\mathcal{A}^\lambda$ and $\mathcal{I}^\lambda$ be the active and inactive sets for $x(\lambda) = \mathcal{P}(x - \lambda \nabla f(x))$. Let $i \in \mathcal{A}^*$. By Lemma 5.3.1 and continuity of ∇f there is δ_2 such that if $\|e\| < \delta_2$ then

$$(\nabla f(x^* + e))_i (x - x^*)_i > \sigma |x - x^*|_i / 2.$$

Therefore, if

$$\|e\| < \delta_3 < \min(\sigma/2, \delta_2),$$

then $i \in \mathcal{A}^\lambda$ and $(x(\lambda))_i = (x^*)_i$. Hence $\mathcal{A}^* \subset \mathcal{A}^\lambda$.

It remains to prove that $\mathcal{A}^\lambda \subset \mathcal{A}^*$. By definition of $\mathcal{P}$ we have

$$\|\mathcal{P}(x) - \mathcal{P}(y)\| \le \|x - y\|$$

for all $x, y \in R^N$. Continuity of $\nabla^2 f$ implies that ∇f is Lipschitz continuous. We let L denote the Lipschitz constant of ∇f in Ω. By stationarity and Theorem 5.2.4,

$$x^* = x^*(\lambda) = \mathcal{P}(x^* - \lambda \nabla f(x^*)),$$

and, therefore,

$$\begin{aligned} \|x^* - x(\lambda)\| &= \|\mathcal{P}(x^* - \lambda \nabla f(x^*)) - \mathcal{P}(x - \lambda \nabla f(x))\| \\ &\le \|e\| + \lambda \|\nabla f(x^*) - \nabla f(x)\| \le (1 + L\lambda)\|e\|. \end{aligned} \tag{5.14}$$

If there is $i \in \mathcal{A}^\lambda \cap \mathcal{I}^*$ then we must have

$$\|x^* - x(\lambda)\| \ge \delta_1 = \min_{i \in \mathcal{I}^*}\{(U_i - x^*), (x^* - L_i)\}. \tag{5.15}$$

However, if

$$\|e\| < \delta_4 = \min(\delta_3, \delta_1/(1+L))$$

then (5.14) implies that (5.15) cannot hold. This completes the proof. ☐

THEOREM 5.4.2. *Let f be twice continuously differentiable on Ω and let x^* be a nondegenerate stationary point for problem (5.4). Assume that sufficient conditions hold at x^*. Then there are δ and M such that if $\|e\| < \delta$ and $\mathcal{A}(x) = \mathcal{A}(x^*)$ then*

$$\|e\|/M \le \|x - x(1)\| \le M\|e\|. \tag{5.16}$$

Proof. Again we let L denote the Lipschitz constant of ∇f in Ω and

$$\mathcal{A}^* = \mathcal{A}(x^*), \mathcal{I}^* = \mathcal{I}(x^*), \mathcal{A} = \mathcal{A}(x), \text{ and } \mathcal{I} = \mathcal{I}(x).$$

Using stationarity we obtain

$$\begin{aligned} \|x - x(1)\| &= \|e - (x(1) - x^*(1))\| \\ &\leq \|e\| + \|\mathcal{P}(x - \nabla f(x)) - \mathcal{P}(x^* - \nabla f(x^*))\| \\ &\leq 2\|e\| + \|\nabla f(x) - \nabla f(x^*)\| \leq (2 + L)\|e\|. \end{aligned}$$

Hence, the right inequality in (5.16) holds.

To verify the left inequality in (5.16) we apply Lemma 5.4.1. Let δ_1 be such that $\|e\| < \delta_1$ implies that the conclusions of Lemma 5.4.1 hold for $\lambda = 1$. The lemma implies that

$$(x - x(1))_i = \begin{cases} (\nabla f(x))_i, & i \in \mathcal{I}^*, \\ (e)_i = 0, & i \in \mathcal{A}^*. \end{cases} \tag{5.17}$$

The remaining case is if $i \in \mathcal{I} = \mathcal{I}^*$. The sufficiency conditions imply that there is $\mu > 0$ such that

$$u^T \mathcal{P}_{\mathcal{I}^*} \nabla^2 f(x^*) \mathcal{P}_{\mathcal{I}^*} u \geq \mu \|\mathcal{P}_{\mathcal{I}^*} u\|^2$$

for all $u \in R^N$. Hence, there is δ_2 so that if $\|e\| < \delta_2$ then

$$u^T \mathcal{P}_{\mathcal{I}^*} \nabla^2 f(x) \mathcal{P}_{\mathcal{I}^*} u \geq \mu \|\mathcal{P}_{\mathcal{I}^*} u\|^2 / 2$$

for all $u \in R^N$.

Therefore, since $e = \mathcal{P}_{\mathcal{I}^*} e$,

$$\begin{aligned} \|\mathcal{P}_{\mathcal{I}^*}(x - x(1))\|^2 &= \int_0^1 e^T \mathcal{P}_{\mathcal{I}^*} \nabla^2 f(x^* + te) e \, dt \\ &= \int_0^1 e^T \mathcal{P}_{\mathcal{I}^*} \nabla^2 f(x^* + te) \mathcal{P}_{\mathcal{I}^*} e \, dt \\ &\geq \mu \|\mathcal{P}_{\mathcal{I}}^* e\|^2 / 2. \end{aligned}$$

Therefore, $\|x - x(1)\| \geq \min(1, \sqrt{\mu/2}) \|e\|$ and setting $M = \max\{2 + L, 1, \sqrt{2/\mu}\}$ completes the proof. □

Following the unconstrained case, we formulate a termination criterion based on relative and absolute reductions in the *measure of stationarity* $\|x - x(1)\|$. Given $r_0 = \|x_0 - x_0(1)\|$ and relative and absolute tolerances τ_r and τ_a the termination criterion for Algorithm `gradproj` is

$$\|x - x(1)\| \leq \tau_a + \tau_r r_0. \tag{5.18}$$

5.4.2 Convergence Analysis

The convergence analysis is more complicated than that for the steepest descent algorithm because of the care that must be taken with the constraints. Our analysis begins with several preliminary lemmas.

LEMMA 5.4.3. *For all* $x, y \in \Omega$,

$$(y - x(\lambda))^T (x(\lambda) - x + \lambda \nabla f(x)) \geq 0. \tag{5.19}$$

Proof. By definition of $\mathcal{P}$

$$\|x(\lambda) - x + \lambda \nabla f(x)\| \leq \|y - x + \lambda \nabla f(x)\|$$

for all $y \in \Omega$. Hence $t = 0$ is a local minimum for

$$\phi(t) = \|(1-t)x(\lambda) + ty - x + \lambda \nabla f(x)\|^2/2$$

and, therefore,

$$0 \leq \phi'(0) = (y - x(\lambda))^T(x(\lambda) - x + \lambda \nabla f(x))$$

as asserted. □

We will most often use the equivalent form of (5.19)

$$(x - x(\lambda))^T(y - x(\lambda)) \leq \lambda \nabla f(x)^T(y - x(\lambda)). \tag{5.20}$$

Setting $y = x$ in (5.20), we state Corollary 5.4.4.

COROLLARY 5.4.4. *For all $x \in \Omega$ and $\lambda \geq 0$,*

$$\|x - x(\lambda)\|^2 \leq \lambda \nabla f(x)^T(x - x(\lambda)). \tag{5.21}$$

An important result in any line search analysis is that the steplengths remain bounded away from 0.

THEOREM 5.4.5. *Assume that ∇f is Lipschitz continuous with Lipschitz constant L. Let $x \in \Omega$. Then the sufficient decrease condition* (5.13) *holds for all λ such that*

$$0 < \lambda \leq \frac{2(1-\alpha)}{L}. \tag{5.22}$$

Proof. We begin with the fundamental theorem of calculus. Setting $y = x - x(\lambda)$ we have

$$f(x-y) - f(x) = f(x(\lambda)) - f(x) = -\int_0^1 \nabla f(x - ty)^T y \, dt.$$

Hence,

$$\begin{aligned} f(x(\lambda)) &= f(x) + \nabla f(x)^T(x(\lambda) - x) \\ &\quad - \int_0^1 (\nabla f(x - ty) - \nabla f(x))^T y \, dt. \end{aligned} \tag{5.23}$$

Rearranging terms in (5.23) gives

$$\lambda(f(x) - f(x(\lambda))) = \lambda \nabla f(x)^T(x - x(\lambda)) + \lambda E, \tag{5.24}$$

where

$$E = \int_0^1 (\nabla f(x - ty) - \nabla f(x))^T y \, dt$$

and hence

$$\|E\| \leq L\|x - x(\lambda)\|^2/2.$$

So

$$\lambda(f(x) - f(x(\lambda))) \geq \lambda \nabla f(x)^T (x - x(\lambda)) - \lambda L \|x - x(\lambda)\|^2/2. \tag{5.25}$$

Therefore, using Corollary 5.4.4 we obtain

$$\lambda(f(x) - f(x(\lambda))) \geq (1 - \lambda L/2)\|x - x(\lambda)\|^2$$

which completes the proof. □

The consequence for the Armijo rule is that the line search will terminate when

$$\beta^m \leq \frac{2(1-\alpha)}{L} \leq \beta^{m-1}$$

if not before. Hence, a lower bound for the steplengths is

$$\bar{\lambda} = \frac{2\beta(1-\alpha)}{L}. \tag{5.26}$$

THEOREM 5.4.6. *Assume that ∇f is Lipschitz continuous with Lipschitz constant L. Let $\{x_n\}$ be the sequence generated by the gradient projection method. Then every limit point of the sequence is a stationary point.*

Proof. Since the sequence $\{f(x_n)\}$ is decreasing and f is bounded from below on Ω, $f(x_n)$ has a limit f^*. The sufficient decrease condition, as in the proof of Theorem 3.2.4, and (5.26) imply that

$$\|x_n - x_{n+1}\|^2 \leq \lambda(f(x_n) - f(x_{n+1}))/\alpha \leq (f(x_n) - f(x_{n+1}))/\alpha \to 0$$

as $n \to \infty$.

Now let $y \in \Omega$ and $n \geq 0$. By (5.20) we have

$$\begin{aligned} \nabla f(x_n)^T (x_n - y) &= \nabla f(x_n)^T (x_{n+1} - y) + \nabla f(x_n)^T (x_n - x_{n+1}) \\ &\leq \lambda_n^{-1} (x_n - x_{n+1})^T (x_{n+1} - y) + \nabla f(x_n)^T (x_n - x_{n+1}). \end{aligned}$$

Therefore, by (5.26),

$$\begin{aligned} \nabla f(x_n)^T (x_n - y) &\leq \|x_n - x_{n+1}\|(\lambda_n^{-1}\|x_{n+1} - y\| + \|\nabla f(x_n)\|), \\ \nabla f(x_n)^T (x_n - y) &\leq \|x_n - x_{n+1}\|(\bar{\lambda}^{-1}\|x_{n+1} - y\| + \|\nabla f(x_n)\|). \end{aligned} \tag{5.27}$$

If $x_{n_l} \to x^*$ is a convergence subsequence of $\{x_n\}$, then we may take limits in (5.27) as $l \to \infty$ and complete the proof. □

5.4.3 Identification of the Active Set

The gradient projection iteration has the remarkable property that if it converges to a nondegenerate local minimizer, then the active set $\mathcal{A}^n$ of x_n is the same as $\mathcal{A}^*$ after only finitely many iterations.

THEOREM 5.4.7. *Assume that f is Lipschitz continuously differentiable and that the gradient projection iterates $\{x_n\}$ converge to a nondegenerate local minimizer x^*. Then there is n_0 such that $\mathcal{A}(x_n) = \mathcal{A}(x^*)$ for all $n \geq n_0$.*

Proof. Let $\bar{\lambda}$ be the lower bound for the steplength. Let δ be such that the conclusions of Lemma 5.4.1 hold for $\lambda = \bar{\lambda}$ (and hence for all $\lambda \geq \bar{\lambda}$). Let n_0 be such that $\|e_n\| < \delta$ for all $n \geq n_0 - 1$ and the proof is complete. □

5.4.4 A Proof of Theorem 5.2.4

We close this section with a proof of Theorem 5.2.4. We define a nonsmooth function

$$F(x) = x - \mathcal{P}(x - \nabla f(x)). \tag{5.28}$$

Using (5.12),

$$F(x) = x - x(1).$$

We now prove Theorem 5.2.4.

Proof. Corollary 5.4.4 states that

$$\|x^* - x^*(\lambda)\|^2 \le \lambda \nabla f(x^*)^T (x^* - x^*(\lambda)).$$

If we set $x = x^*(\lambda)$ in the definition of stationarity (5.7) we have

$$\nabla f(x^*)^T (x^* - x^*(\lambda)) \le 0$$

and hence $x^* = x^*(\lambda)$.

Conversely assume that $x^* = x^*(\lambda)$ for all $\lambda > 0$. This implies that x^* is left invariant by the gradient projection iteration and is therefore a stationary point. □

By setting $\lambda = 1$ we obtain a simple consequence of Theorem 5.2.4.

COROLLARY 5.4.8. *Let f be a Lipschitz continuously differentiable function on Ω. Then if x^* is stationary then $F(x^*) = 0$.*

5.5 Superlinear Convergence

Once the gradient projection iteration has identified the active constraints, $\mathcal{P}_{\mathcal{A}(x^*)} x^*$ is known. At that point the minimization problem for $\mathcal{P}_{\mathcal{I}} x^*$ is unconstrained and, in principal, any superlinearly convergent method for unconstrained optimization could then be used.

The problem with this idea is, of course, that determining when the active set has been identified is possible only after the problem has been solved and an error in estimating the active set can have devastating effects on convergence. In this section we discuss two approaches: one, based on Newton's method, is presented only as a local method; the other is a BFGS–Armijo method similar to Algorithm `bfgsopt`.

We will begin with the development of the local theory for the projected Newton method [19]. This analysis illustrates the important problem of estimation of the active set. As with the unconstrained minimization problem, the possibility of negative curvature makes this method difficult to globalize (but see §5.6 for pointers to the literature on trust region methods). Following the approach in §4.2 we describe a projected BFGS–Armijo scheme in §5.5.3.

5.5.1 The Scaled Gradient Projection Algorithm

One might think that the theory developed in §5.4 applies equally well to iterations of the form

$$x_+ = \mathcal{P}(x_c - \lambda H_c^{-1} \nabla f(x_c))$$

where H_c is spd. This is not the case as the following simple example illustrates. Let $N = 2$, $L_i = 0$, and $U_i = 1$ for all i. Let

$$f(x) = \|x - (-1, 1/2)^T\|^2/2;$$

then the only local minimizer for (5.3) is $x^* = (0, 1/2)^T$. Let $x_c = (0, 0)$ (not a local minimizer!); then $\nabla f(x_c) = (1, -1/2)^T$. If

$$H_c^{-1} = \begin{pmatrix} 2 & 1 \\ 1 & 2 \end{pmatrix}$$

then H_c^{-1} (and hence H_c) is spd, and

$$H_c^{-1}\nabla f(x_c) = \begin{pmatrix} 2 & 1 \\ 1 & 2 \end{pmatrix} \begin{pmatrix} 1 \\ -1/2 \end{pmatrix} = \begin{pmatrix} 3/2 \\ 0 \end{pmatrix}.$$

Therefore, for all $\lambda > 0$,

$$x_c(\lambda) = \mathcal{P}\left(\begin{pmatrix} 0 \\ 0 \end{pmatrix} - \lambda H_c^{-1} \begin{pmatrix} 1 \\ -1/2 \end{pmatrix}\right) = \mathcal{P}\begin{pmatrix} -3\lambda/2 \\ 0 \end{pmatrix} = \begin{pmatrix} 0 \\ 0 \end{pmatrix} = x_c.$$

The reason that $x_c(\lambda) = x_c$ for all $\lambda > 0$ is that the search direction for the unconstrained problem has been rotated by H_c^{-1} to be orthogonal to the direction of decrease in the inactive directions for the constrained problem. Hence, unlike the constrained case, positivity of the model Hessian is not sufficient and we must be able to estimate the active set and model the reduced Hessian (rather than the Hessian) if we expect to improve convergence.

The solution proposed in [19] is to underestimate the inactive set in a careful way and therefore maintain a useful spd approximate reduced Hessian. For $x \in \Omega$ and

$$0 \le \epsilon < \min(U_i - L_i)/2,$$

we define $\mathcal{A}^\epsilon(x)$, the *ϵ-active set* at x, by

(5.29) $$\mathcal{A}^\epsilon(x) = \{i \,|\, U_i - (x)_i \le \epsilon \text{ or } (x)_i - L_i \le \epsilon\}.$$

And let $\mathcal{I}^\epsilon(x)$, the *ϵ-inactive set*, be the complement of $\mathcal{A}^\epsilon(x)$.

Given $0 \le \epsilon_c < \min(U_i - L_i)/2$, x_c, and an spd matrix H_c, we model the reduced Hessian with $\mathcal{R}_c$, the matrix with entries

$$\mathcal{R}_c = \mathcal{P}_{\mathcal{A}^{\epsilon_c}(x_c)} + \mathcal{P}_{\mathcal{I}^{\epsilon_c}(x_c)} H_c \mathcal{P}_{\mathcal{I}^{\epsilon_c}(x_c)} = \begin{cases} \delta_{ij} & \text{if } i \in \mathcal{A}^{\epsilon_c}(x_c) \text{ or } j \in \mathcal{A}^{\epsilon_c}(x_c), \\ (H_c)_{ij} & \text{otherwise.} \end{cases}$$

(5.30)

When the explicit dependence on x_c, ϵ_c, and H_c is important we will write

$$\mathcal{R}(x_c, \epsilon_c, H_c).$$

So, for example,

$$\nabla_R^2 f(x_c) = \mathcal{R}(x_c, 0, \nabla^2 f(x_c)).$$

Given $0 < \epsilon < \min(U_i - L_i)/2$ and an spd H, define

$$x^{H,\epsilon}(\lambda) = \mathcal{P}(x - \lambda \mathcal{R}(x, \epsilon, H)^{-1} \nabla f(x)).$$

It requires proof that

$$f(x^{H,\epsilon}(\lambda)) < f(x)$$

for λ sufficiently small. We prove more and show that the *sufficient decrease* condition

(5.31) $$f(x^{H,\epsilon}(\lambda)) - f(x) \le -\alpha \nabla f(x)^T (x - x^{H,\epsilon}(\lambda))$$

holds for sufficiently small λ.

LEMMA 5.5.1. *Let $x \in \Omega$, $0 < \epsilon < \min(U_i - L_i)/2$, and H be spd with smallest and largest eigenvalues $0 < \lambda_s \le \lambda_l$. Let ∇f be Lipschitz continuous on Ω with Lipschitz constant L. Then there is $\bar\lambda(\epsilon, H)$ such that* (5.31) *holds for all*

$$\lambda \le \bar\lambda(\epsilon, H). \tag{5.32}$$

Proof. The proof will show that

$$\nabla f(x)^T(x - x^{H,\epsilon}(\lambda)) \ge \lambda_l^{-1}\nabla f(x)^T(x - x(\lambda))$$

and then use the method of proof from Theorem 5.4.5. We do this by writing

$$\nabla f(x)^T(x - x^{H,\epsilon}(\lambda)) = (\mathcal{P}_{\mathcal{A}^\epsilon(x)}\nabla f(x))^T(x - x^{H,\epsilon}(\lambda)) + (\mathcal{P}_{\mathcal{I}^\epsilon(x)}\nabla f(x))^T(x - x^{H,\epsilon}(\lambda))$$

and considering the two terms on the right side separately.

We begin by looking at $(\mathcal{P}_{\mathcal{A}^\epsilon(x)}\nabla f(x))^T(x - x^{H,\epsilon}(\lambda))$. Note that

$$(x^{H,\epsilon}(\lambda))_i = (x(\lambda))_i \text{ for } i \in \mathcal{A}^\epsilon(x)$$

and, therefore,

$$(\mathcal{P}_{\mathcal{A}^\epsilon(x)}\nabla f(x))^T(x - x^{H,\epsilon}(\lambda)) = (\mathcal{P}_{\mathcal{A}^\epsilon(x)}\nabla f(x))^T(x - x(\lambda)). \tag{5.33}$$

We will need to show that

$$(\mathcal{P}_{\mathcal{A}^\epsilon(x)}\nabla f(x))^T(x - x(\lambda)) \ge 0. \tag{5.34}$$

Now assume that

$$\lambda < \bar\lambda_1 = \frac{\min(U_i - L_i)}{2\max_{x\in\Omega}\|\nabla f(x)\|_\infty}. \tag{5.35}$$

Since $\mathcal{A}(x) \subset \mathcal{A}^\epsilon(x)$ we can investigate the contributions of $\mathcal{A}(x)$ and $\mathcal{A}^\epsilon(x) \cap \mathcal{I}(x)$ separately.

If $i \in \mathcal{A}(x)$ then (5.35) implies that either $(x - x(\lambda))_i = \lambda(\nabla f(x))_i$ or $(x - x(\lambda))_i = 0$. In either case $(x - x(\lambda))_i(\nabla f(x))_i \ge 0$. If $i \in \mathcal{A}^\epsilon(x) \cap \mathcal{I}(x)$ and $(x - x(\lambda))_i \ne \lambda(\nabla f(x))_i$, then it must be the case that $i \in \mathcal{A}(x(\lambda))$ and therefore we must still have $(x - x(\lambda))_i(\nabla f(x))_i \ge 0$. Hence (5.34) holds.

Now if $i \in \mathcal{I}^\epsilon(x)$ then, by definition,

$$L_i + \epsilon \le (x)_i \le U_i - \epsilon$$

and, hence, if

$$\lambda \le \bar\lambda_2 = \frac{\epsilon}{\max_{x\in\Omega}\|\mathcal{R}(x, \epsilon, H)^{-1}\nabla f(x)\|_\infty} \tag{5.36}$$

then i is in the inactive set for both $x^{H,\epsilon}(\lambda)$ and $x(\lambda)$. Therefore, if (5.36) holds then

$$\begin{aligned}
(\mathcal{P}_{\mathcal{I}^\epsilon(x)}\nabla f(x))^T(x - x^{H,\epsilon}(\lambda)) &= \lambda(\mathcal{P}_{\mathcal{I}^\epsilon(x)}\nabla f(x))^T H^{-1}\mathcal{P}_{\mathcal{I}^\epsilon(x)}\nabla f(x) \\
&\ge \lambda_l^{-1}\lambda^{-1}\|\mathcal{P}_{\mathcal{I}^\epsilon(x)}(x - x(\lambda))\|^2 \qquad (5.37) \\
&= \lambda_l^{-1}(\mathcal{P}_{\mathcal{I}^\epsilon(x)}\nabla f(x))^T(x - x(\lambda)).
\end{aligned}$$

Hence, using Corollary 5.4.4, (5.33), (5.34), and (5.37), we obtain

$$\begin{aligned}\nabla f(x)^T(x - x^{H,\epsilon}(\lambda)) &= (\mathcal{P}_{\mathcal{A}^\epsilon(x)}\nabla f(x))^T(x - x^{H,\epsilon}(\lambda)) \\ &\quad +(\mathcal{P}_{\mathcal{I}^\epsilon(x)}\nabla f(x))^T(x - x^{H,\epsilon}(\lambda)) \\ &\geq (\mathcal{P}_{\mathcal{A}^\epsilon(x)}\nabla f(x))^T(x - x(\lambda)) + \lambda_l^{-1}(\mathcal{P}_{\mathcal{I}^\epsilon(x)}\nabla f(x))^T(x - x(\lambda)) \\ &\geq \min(1, \lambda_l^{-1})\nabla f(x)^T(x - x(\lambda)) \geq \frac{\min(1,\lambda_l^{-1})}{\lambda}\|x - x(\lambda)\|^2.\end{aligned} \tag{5.38}$$

The remainder of the proof is almost identical to that for Theorem 5.4.5. The fundamental theorem of calculus and the Lipschitz continuity assumption imply that

$$f(x^{H,\epsilon}(\lambda)) - f(x) \leq -\nabla f(x)^T(x - x^{H,\epsilon}(\lambda)) + L\|x - x^{H,\epsilon}(\lambda)\|^2.$$

We apply (5.38) and obtain

$$f(x^{H,\epsilon}(\lambda)) - f(x) \leq -(1 - L\lambda \max(1, \lambda_l))\nabla f(x)^T(x - x^{H,\epsilon}(\lambda)),$$

which implies (5.31) if $1 - L\lambda\max(1, \lambda_l) \geq \alpha$ which will follow from

$$\lambda \leq \bar{\lambda}_3 = \frac{(1-\alpha)}{\max(1,\lambda_l)L}. \tag{5.39}$$

This completes the proof with $\bar{\lambda} = \min(\bar{\lambda}_1, \bar{\lambda}_2, \bar{\lambda}_3)$. □

An algorithm based on these ideas is the *scaled gradient projection* algorithm. The name comes from the *scaling matrix H* that is used to computed the direction. The inputs are the initial iterate, the vectors of upper and lower bounds u and l, the relative-absolute residual tolerance vector $\tau = (\tau_r, \tau_a)$, and a limit on the number of iterations. Left unstated in the algorithmic description are the manner in which the parameter ϵ is computed and the way in which the approximate Hessians are constructed.

ALGORITHM 5.5.1. `sgradproj`$(x, f, \tau, nmax)$

1. *For* $n = 1, \dots, nmax$
   (a) *Compute* f *and* ∇f; *test for termination using* (5.18).
   (b) *Compute* ϵ *and an spd* H.
   (c) *Solve*
   $$\mathcal{R}(x, \epsilon, H_c)d = -\nabla f(x_c).$$
   (d) *Find the least integer* m *such that* (5.13) *holds for* $\lambda = \beta^m$.
   (e) $x = x(\lambda)$.
2. *If* $n = nmax$ *and the termination test is failed, signal failure.*

If our model reduced Hessians remain uniformly positive definite, a global convergence result completely analogous to Theorem 3.2.4 holds.

THEOREM 5.5.2. *Let* ∇f *be Lipschitz continuous with Lipschitz constant L. Assume that the matrices* H_n *are symmetric positive definite and that there are* $\bar{\kappa}$ *and* λ_l *such that* $\kappa(H_n) \leq \bar{\kappa}$, *and* $\|H_n\| \leq \lambda_l$ *for all* n. *Assume that there is* $\bar{\epsilon} > 0$ *such that* $\bar{\epsilon} \leq \epsilon_n < \min(U_i - L_i)/2$ *for all* n.

Then

$$\lim_{n\to\infty} \|x_n - x_n(1)\| = 0, \tag{5.40}$$

and hence any limit point of the sequence of iterates produced by Algorithm `sgradproj` *is a stationary point.*

In particular, if $x_{n_l} \to x^*$ *is any convergent subsequence of* $\{x_n\}$*, then* $x^* = x^*(1)$*. If* x_n *converges to a nondegenerate local minimizer* x^**, then the active set of* x_n *is the same as that of* x^* *after finitely many iterations.*

Proof. With Lemma 5.5.1 and its proof in hand, the proof follows the outline of the proof of Theorem 5.4.6. We invite the reader to work through it in exercise 5.8.3. □

5.5.2 The Projected Newton Method

The requirement in the hypothesis of Theorem 5.5.2 that the sequence $\{\epsilon_n\}$ be bounded away from zero is used to guarantee that the steplengths λ_n are bounded away from zero. This is needed because ϵ appears in the numerator in (5.36). However, once the active set has been identified and one is near enough to a nondegenerate local minimizer for the reduced Hessians to be spd, one is solving an unconstrained problem. Moreover, once near enough to that minimizer, the convergence theory for Newton's method will hold. Then one can, in principle, set $\epsilon_n = 0$ and the iteration will be q-quadratically convergent. In this section we discuss an approach from [19] for making a transition from the globally convergent regime described in Theorem 5.5.2 to the locally convergent setting where Newton's method converges rapidly.

If the initial iterate x_0 is sufficiently near a nondegenerate local minimizer x^* and we take

$$H_n = \nabla_R^2 f(x_n)$$

in Algorithm `sgradproj`, then the resulting *projected Newton method* will take full steps (i.e., $\lambda = 1$) and, if ϵ_n is chosen with care, converge q-quadratically to x^*.

A specific form of the recommendation from [19], which we use here, is

$$\epsilon_n = \min(\|x_n - x_n(1)\|, \min(U_i - L_i)/2). \tag{5.41}$$

Note that while x_n is far from a stationary point and the reduced Hessian is spd, then ϵ_n will be bounded away from zero and Theorem 5.5.2 will be applicable. The convergence result is like Theorem 2.3.3 for local convergence but makes the strong assumption that H_n is spd (valid near x^*, of course) in order to get a global result.

Algorithm `projnewt` is the formal description of the projected Newton algorithm. It is a bit more than just a specific instance of Algorithm `gradproj`. Keep in mind that if the initial iterate is far from x^* and the reduced Hessian is not spd, then the line search (and hence the entire iteration) may fail. The algorithm tests for this. This possibility of indefiniteness is the weakness in any line search method that uses $\nabla^2 f$ when far from the minimizer. The inputs to Algorithm `projnewt` are the same as those for Algorithm `gradproj`. The algorithm exploits the fact that

$$\mathcal{R}(x, \epsilon, \nabla_R^2 f(x)) = \mathcal{R}(x, \epsilon, \nabla^2 f(x)) \tag{5.42}$$

which follows from $\mathcal{A}(x) \subset \mathcal{A}^\epsilon(x)$.

ALGORITHM 5.5.2. `projnewt`$(x, f, \tau, nmax)$

1. *For* $n = 1, \dots, nmax$
   (a) *Compute* f *and* ∇f*; test for termination using* (5.18).
   (b) *Set* $\epsilon = \|x - x(1)\|$.

(c) *Compute and factor* $\mathcal{R} = \mathcal{R}(x, \epsilon, \nabla^2_R f(x))$. *If* $\mathcal{R}$ *is not spd, terminate with a failure message.*

(d) *Solve* $\mathcal{R}d = -\nabla f(x_c)$.

(e) *Find the least integer* m *such that* (5.13) *holds for* $\lambda = \beta^m$.

(f) $x = x(\lambda)$.

2. *If* $n = nmax$ *and the termination test is failed, signal failure.*

THEOREM 5.5.3. *Let* x^* *be a nondegenerate local minimizer. Then if* x_0 *is sufficiently near to* x^* *and* $\mathcal{A}(x_0) = \mathcal{A}(x^*)$ *then the projected Newton iteration, with* $\epsilon_n = \|x_n - x_n(1)\|$, *will converge q-quadratically to* x^*.

Proof. Our assumption that the active set has been identified, i.e.,

$$\mathcal{A}(x_c) = \mathcal{A}(x_+) = \mathcal{A}(x^*),$$

implies that

$$\mathcal{P}_{\mathcal{A}(x_c)} e_c = \mathcal{P}_{\mathcal{A}(x_c)} e_+ = 0.$$

Hence, we need only estimate $\mathcal{P}_{\mathcal{I}(x_c)} e_+$ to prove the result.

Let

$$\delta^* = \min_{i \in \mathcal{I}(x^*)} (|(x)_i - U_i|, |(x)_i - L_i|) > 0.$$

We reduce $\|e\|$ if necessary so that

$$\|e\| \le \delta^*/M,$$

where M is the constant in Theorem 5.4.2. We may then apply Theorem 5.4.2 to conclude that both $\epsilon_c < \delta^*$ and $\|e_c\| < \delta^*$. Then any index $i \in \mathcal{A}^{\epsilon_c}(x_c)$ must also be in $\mathcal{A}(x_c) = \mathcal{A}(x^*)$. Hence

$$\mathcal{A}^{\epsilon_c}(x_c) = \mathcal{A}(x_c) = \mathcal{A}(x^*). \tag{5.43}$$

From this we have

$$\mathcal{R}(x_c, \epsilon_c, \nabla^2_R f(x_c)) = \nabla^2_R f(x_c). \tag{5.44}$$

Hence, for $\|e_c\|$ sufficiently small the projected Newton iteration is

$$x_+ = \mathcal{P}(x_c - (\nabla^2_R f(x_c))^{-1} \nabla f(x_c)).$$

By the fundamental theorem of calculus,

$$\nabla f(x_c) = \nabla f(x^*) + \nabla^2 f(x_c) e_c + E_1, \tag{5.45}$$

where

$$E_1 = \int_0^1 (\nabla^2 f(x^* + t e_c) - \nabla^2 f(x_c)) e_c \, dt$$

and hence $\|E_1\| \le K_1 \|e_c\|^2$ for some $K_1 > 0$.

By the necessary conditions,

$$\mathcal{P}_{\mathcal{I}(x)} \nabla f(x^*) = \mathcal{P}_{\mathcal{I}(x^*)} \nabla f(x^*) = 0. \tag{5.46}$$

By the fact that $\mathcal{I}(x_c) = \mathcal{I}(x^*)$, we have the equivalent statements

$$e_c = \mathcal{P}_{\mathcal{I}(x_c)} e_c \text{ and } \mathcal{P}_{\mathcal{A}(x_c)} e_c = 0. \tag{5.47}$$

Therefore, combining (5.45), (5.46), (5.47),

$$
\begin{aligned}
\mathcal{P}_{\mathcal{I}(x_c)}\nabla f(x_c) &= \mathcal{P}_{\mathcal{I}(x_c)}\nabla^2 f(x_c)\mathcal{P}_{\mathcal{I}(x_c)}e_c + \mathcal{P}_{\mathcal{I}(x_c)}E_1 \\
&= \mathcal{P}_{\mathcal{A}(x_c)}e_c + \mathcal{P}_{\mathcal{I}(x_c)}\nabla^2 f(x_c)\mathcal{P}_{\mathcal{I}(x_c)}e_c + \mathcal{P}_{\mathcal{I}(x_c)}E_1 \\
&= \nabla_R^2 f(x_c)e_c + \mathcal{P}_{\mathcal{I}(x_c)}E_1.
\end{aligned} \tag{5.48}
$$

So, by definition of ∇_R^2,

$$
\mathcal{P}_{\mathcal{I}(x_c)}(\nabla_R^2 f(x_c))^{-1}\nabla f(x_c) = (\nabla_R^2 f(x_c))^{-1}\mathcal{P}_{\mathcal{I}(x_c)}\nabla f(x_c) = e_c + E_2,
$$

where $\|E_2\| \le K_2\|e_c\|^2$ for some $K_2 > 0$.

Since $\mathcal{P}_{\mathcal{I}(x_c)}\mathcal{P}w = \mathcal{P}\mathcal{P}_{\mathcal{I}(x_c)}w$ for all $w \in R^N$,

$$
\begin{aligned}
\mathcal{P}_{\mathcal{I}(x_c)}x_+ &= \mathcal{P}_{\mathcal{I}(x_c)}\mathcal{P}(x_c - (\nabla_R^2 f(x_c))^{-1}\nabla f(x_c)) \\
&= \mathcal{P}\mathcal{P}_{\mathcal{I}(x_c)}(x_c - (\nabla_R^2 f(x_c))^{-1}\nabla f(x_c)) = \mathcal{P}(x^* - E_2).
\end{aligned}
$$

Therefore, $\|e_+\| \le K_2\|e_c\|^2$ as asserted. □

5.5.3 A Projected BFGS–Armijo Algorithm

We can apply the structured quasi-Newton updating scheme from §4.3 with

$$
C(x) = \mathcal{P}_{\mathcal{A}^\epsilon(x)} \tag{5.49}
$$

and update an approximation to the part of the model Hessian that acts on the ϵ inactive set. In this way we can hope to maintain a positive definite model reduced Hessian with, say, a BFGS update. So if our model reduced Hessian is

$$
\mathcal{R} = C(x) + A,
$$

we can use (4.42) to update A (with $A_0 = \mathcal{P}_{\mathcal{I}^{\epsilon_0}(x_0)}$, for example), as long as the ϵ active set does not change. If one begins the iteration near a nondegenerate local minimizer with an accurate approximation to the Hessian, then one would expect, based on Theorem 5.5.3, that the active set would remain constant and that the iteration would converge q-superlinearly.

However, if the initial data is far from a local minimizer, the active set can change with each iteration and the update must be designed to account for this. One way to do this is to use a projected form of the BFGS update of A from (4.42),

$$
A_+ = \mathcal{P}_{\mathcal{I}_+}A_c\mathcal{P}_{\mathcal{I}_+} + \frac{y^\# {y^\#}^T}{{y^\#}^T s} - \mathcal{P}_{\mathcal{I}_+}\frac{(A_c s)(A_c s)^T}{s^T A_c s}\mathcal{P}_{\mathcal{I}_+}, \tag{5.50}
$$

with

$$
y^\# = \mathcal{P}_{\mathcal{I}_+}(\nabla f(x_+) - \nabla f(x_c)).
$$

Here $\mathcal{I}_+ = \mathcal{I}^{\epsilon_+}(x_+)$. This update carries as much information as possible from the previous model reduced Hessian while taking care about proper approximation of the active set. As in the unconstrained case, if ${y^\#}^T s \le 0$ we can either skip the update or reinitialize A to $\mathcal{P}_{\mathcal{I}}$.

A is not spd if any constraints are active. However, we can demand that A be symmetric indefinite, and a generalized inverse $A^\dagger$ exists. We have

$$
(\mathcal{P}_{\mathcal{A}} + A)^{-1} = \mathcal{P}_{\mathcal{A}} + A^\dagger. \tag{5.51}
$$

If $\mathcal{A}(x_c) = \mathcal{A}(x_+)$ any of the low-storage methods from Chapter 4 can be used to update $A^\dagger$. In this case we have

$$A_+^\dagger = \left(I - \frac{sy^{\#T}}{y^{\#T}s}\right) A_c^\dagger \left(I - \frac{y^{\#}s^T}{y^{\#T}s}\right) + \frac{ss^T}{y^{\#T}s}. \tag{5.52}$$

Since $s = \mathcal{P}_{\mathcal{I}_+} s$ if $\mathcal{A}(x_+) = \mathcal{A}(x_c)$, we could replace s by $s^{\#} = \mathcal{P}_{\mathcal{I}_+} s$ in (5.52).

If the set of active constraints changes, then (5.52) is no longer true and we cannot replace s by $s^{\#}$. One approach is to store A, update it as a full matrix, and refactor with each nonlinear iteration. This is very costly for even moderately large problems. Another approach, used in [250], is to reinitialize $A = \mathcal{P}_{\mathcal{I}}$ whenever the active set changes. The problem with this is that in the terminal phase of the iteration, when most of the active set has been identified, too much information is lost when A is reinitialized.

In this book we suggest an approach based on the recursive BFGS update that does not discard information corresponding to that part of the inactive set that is not changed. The idea is that even if the active set has changed, we can still maintain an approximate generalized inverse with

$$A_+^\dagger = \left(I - \frac{s^{\#}y^{\#T}}{y^{\#T}s^{\#}}\right) \mathcal{P}_{\mathcal{I}_+} A_c^\dagger \mathcal{P}_{\mathcal{I}_+} \left(I - \frac{y^{\#}s^{\#T}}{y^{\#T}s^{\#}}\right) + \frac{s^{\#}s^{\#T}}{y^{\#T}s^{\#}}. \tag{5.53}$$

The formulation we use in the MATLAB code `bfgsbound` is based on (5.52) and Algorithm `bfgsrec`. Algorithm `bfgsrecb` stores the sequences $\{y_k^{\#}\}$ and $\{s_k^{\#}\}$ and uses Algorithm `bfgsrec` and the projection $\mathcal{P}_{\mathcal{I}_+}$ to update $A^\dagger$ as the iteration progresses. The data are the same as for Algorithm `bfgsrec` with the addition of

$$\mathcal{P}_{\mathcal{I}_n} = \mathcal{P}_{\mathcal{I}^{\epsilon_n}(x_n)}.$$

Note that the sequences $\{y_k^{\#}\}$ and $\{s_k^{\#}\}$ are changed early in the call and then the unconstrained algorithm `bfgsrec` is used to do most of the work.

ALGORITHM 5.5.3. `bfgsrecb`$(n, \{s_k^{\#}\}, \{y_k^{\#}\}, A_0^\dagger, d, \mathcal{P}_{\mathcal{I}_n})$

1. $d = \mathcal{P}_{\mathcal{I}_n} d$.
2. *If* $n = 0$, $d = A_0^\dagger d$; *return*
3. $\alpha = s^{\#T}_{n-1} d / y_{n-1}^{\# \, T} s^{\#}$; $d = d - \alpha y_{n-1}^{\#}$
4. *call* `bfgsrec`$(n-1, \{s^{\#}{}_k\}, \{y_k^{\#}\}, A_0^\dagger, d)$
5. $d = d + (\alpha - (y_{n-1}^{\# \, T} d / y_{n-1}^{\# \, T} s^{\#}{}_{n-1})) s^{\#}{}_{n-1}$
6. $d = \mathcal{P}_{\mathcal{I}_n} d$.

The projected BFGS–Armijo algorithm that we used in the example problems in §5.7 is based on Algorithm `bfgsrecb`. Note that we reinitialize ns to zero (i.e., reinitialize A to $\mathcal{P}_{\mathcal{I}}$) when $y_{ns}^{\# \, T} s \le 0$. We found experimentally that this was better than skipping the update.

ALGORITHM 5.5.4. `bfgsoptb`(x, f, τ, u, l)

1. $ns = n = 0$; $pg_0 = pg = x - \mathcal{P}(x - \nabla f(x))$
2. $\epsilon = \min(\min(U_i - L_i)/2, \|pg\|)$; $\mathcal{A} = \mathcal{A}^\epsilon(x)$; $\mathcal{I} = \mathcal{I}^\epsilon(x)$; $A_0 = \mathcal{P}_{\mathcal{I}}$
3. *While* $\|pg\| \le \tau_a + \tau_r \|pg_0\|$

(a) $d = -\nabla f(x)$; *Call* `bfgsrecb`$(ns, \{s_k^\#\}, \{y_k^\#\}, A_0^\dagger, d, \mathcal{P}_\mathcal{I})$

(b) $d = -\mathcal{P}_\mathcal{A}\nabla f(x) + d$

(c) *Find the least integer* m *such that* (5.13) *holds for* $\lambda = \beta^m$. *Set* $s_{ns}^\# = \mathcal{P}_\mathcal{I}(x(\lambda) - x)$

(d) $xp = x(\lambda)$; $y = \nabla f(xp) - \nabla f(x)$; x=xp; $y_{ns}^\# = \mathcal{P}_\mathcal{I}(\nabla f(xp) - \nabla f(x))$

(e) *If* ${y_{ns}^\#}^T s_n^\# s > 0$ *then* $ns = ns + 1$, *else* $ns = 0$

(f) $x = xp$; $pg = x - \mathcal{P}(x - \nabla f(x))$

(g) $\epsilon = \min(\min(U_i - L_i)/2, \|pg\|)$; $\mathcal{A} = \mathcal{A}^\epsilon(x)$; $\mathcal{I} = \mathcal{I}^\epsilon(x)$

(h) $n = n + 1$

Theorem 4.1.3 can be applied directly once the active set has been identified and a good initial approximation to the reduced Hessian is available. The reader is invited to construct the (easy!) proof in exercise 5.8.6.

THEOREM 5.5.4. *Let x^* be a nondegenerate local minimizer. Then if x_0 is sufficiently near to x^*, $\mathcal{A}(x_0) = \mathcal{A}(x^*)$, and A_0 sufficiently near to $\mathcal{P}_{\mathcal{I}(x^*)}\nabla^2 f(x^*)\mathcal{P}_{\mathcal{I}(x^*)}$, then the projected BFGS iteration, with $\epsilon_n = \|x_n - x_n(1)\|$, will converge q-superlinearly to x^*.*

A global convergence result for this projected BFGS algorithm can be derived by combining Theorems 5.5.2 and 4.1.9.

THEOREM 5.5.5. *Let ∇f be Lipschitz continuous on Ω. Assume that the matrices H_n are constructed with the projected BFGS method* (5.50) *and satisfy the assumptions of Theorem* 5.5.2. *Then* (5.40) *and the conclusions of Theorem* 5.5.2 *hold.*

Moreover, if x^* is a nondegenerate local minimizer such that there is n_0 such that $\mathcal{A}(x_n) = \mathcal{A}(x^*)$ for all $n \geq n_0$, H_{n_0} is spd, and the set

$$D = \{x \mid f(x) \leq f(x_{n_0}) \text{ and } \mathcal{A}(x) = \mathcal{A}(x^*)\}$$

is convex, then the projected BFGS–Armijo algorithm converges q-superlinearly to x^*.

5.6 Other Approaches

Our simple projected-BFGS method is effective for small to medium sized problems and for very large problems that are discretizations of infinite-dimensional problems that have the appropriate compactness properties. The example in §4.4.2 nicely illustrates this point. For other kinds of large problems, however, more elaborate methods are needed, and we present some pointers to the literature in this section.

The limited memory BFGS method for unconstrained problems described in [44] and [176] has also been extended to bound constrained problems [42], [280]. More general work on line search methods for bound constrained problems can be found in [47], [194], and [42].

Very general theories have been developed for convergence of trust region methods for bound constrained problems. The notion of Cauchy decrease can, for example, be replaced by the decrease from a gradient projection step for the quadratic model [191], [259], [66]. One could look for minima of the quadratic model along the projection path [63], [64], or attempt to project the solution of an unconstrained model using the reduced Hessian [162].

A completely different approach can be based on interior point methods. This is an active research area and the algorithms are not, at least at this moment, easy to implement or analyze. This line of research began with [57] and [58]. We refer the reader to [86] and [266] for more recent accounts of this aspect of the field and to [140] and [79] for some applications to control problems and an account of the difficulties in infinite dimensions.

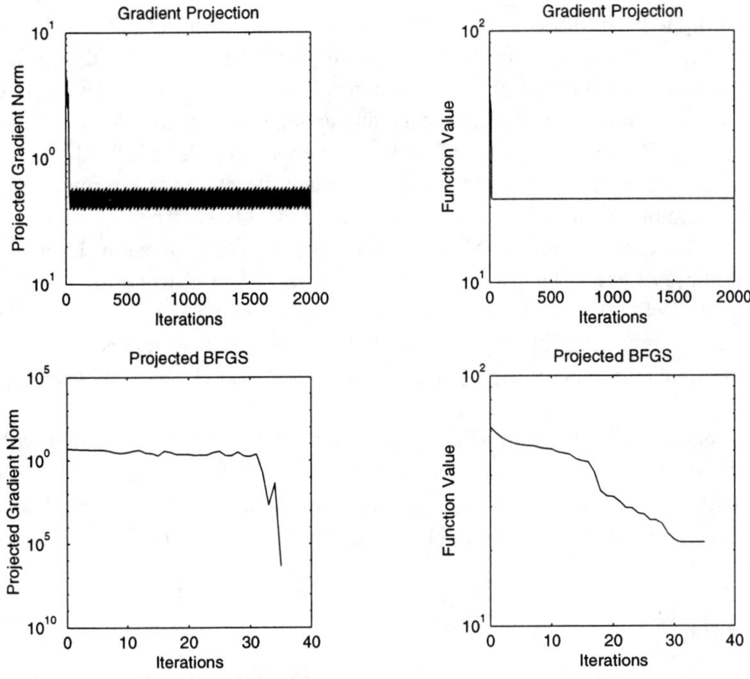

Figure 5.1: *Solution to Constrained Parameter ID Problem*

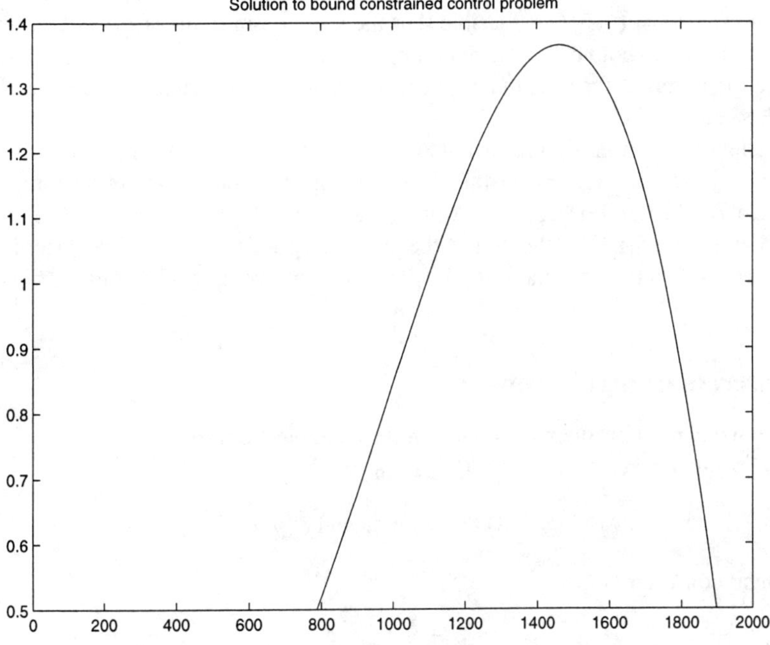

Figure 5.2: *Solution to Discrete Control Problem: First Example*

5.6.1 Infinite-Dimensional Problems

The results in this part of this book do not extend in a direct way to infinite-dimensional problems. One reason for this is that often infinite-dimensional problems have countably infinitely many constraints or even a continuum of constraints; hence Ω is not compact in the norm topology of the Banach space in which the problem is posed and appeals to various types of weak continuity must be made (see [122] for an example of such arguments and [122] and [10] for applications). Moreover, identification of an active set in finitely many iterations is not always possible. A more complete account of this issue may be found in [254], [162], [161].

These are not the only complications that can arise in infinite dimension. Even the projected gradient method presents challenges, especially if the minima fail to be nondegenerate in the sense of this book [94], [95]. Convergence behavior for discretized problems can be different from that for the continuous problem [97]. Nonequivalence of norms makes convergence results difficult to formulate and analyze for both line search [96], [254], [98] and trust region [140], [162] methods.

The functional analytic structure of many control problems can be exploited with fast multilevel methods. Both second kind multigrid methods from [138] and variants of the Atkinson–Brakhage method [9], [31] have been applied to fixed point formulations of parabolic boundary control problems in one space dimension [136], [137], [153], [162], [161].

5.7 Examples

The computations in this section were done with the MATLAB code `bfgsbound`. In this code the storage is limited to five pairs of vectors, and $\beta = .1$ was used in the line search.

5.7.1 Parameter ID Problem

We consider the parameter problem from §3.4.1 with bounds $L = (2,0)^T$ and $U = (20,5)^T$. The initial iterate $x_0 = (5,5)^T$ is feasible, but the global minimum of $(1,1)^T$ is not. As one might expect, the lower bound constraint on $(x)_1$ is active at the optimal point $x^* \approx (2, 1.72)^T$. The termination criterion for both the gradient projection and projected BFGS algorithms was $\|u - u(1)\| \le 10^{-6}$.

The gradient projection algorithm failed. While the value of the objective function was correct, the projected gradient norm failed to converge and the active set was not identified. The projected BFGS iteration converged in 35 iterations. One can see the local superlinear convergence in Figure 5.1 from the plot of the projected gradient norms. The cost of the BFGS iteration was 121 function evaluations, 36 gradients, and roughly 5.3 million floating point operations.

5.7.2 Discrete Control Problem

We base the two control problem examples on the example from §1.6.1.

Our first example takes $N = 2000$, $T = 1$, $y_0 = 0$,

$$L(y,u,t) = (y-3)^2 + .1 * u^2, \text{ and } \phi(y,u,t) = uy + t^2,$$

with the bound constraints

$$.5 \le u \le 2,$$

and the initial iterate $u_0 = 2$. We terminated the iteration when $\|u - u(1)\| \le 10^{-5}$. In Figure 5.2 we plot the solution of this problem. Clearly the active set is not empty for the constrained problem.

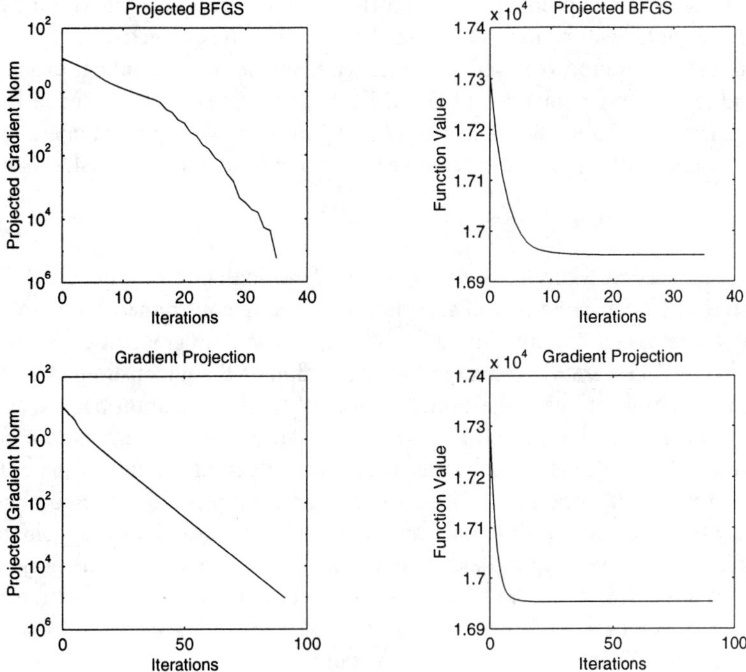

Figure 5.3: *Constrained Discrete Control Problem* I

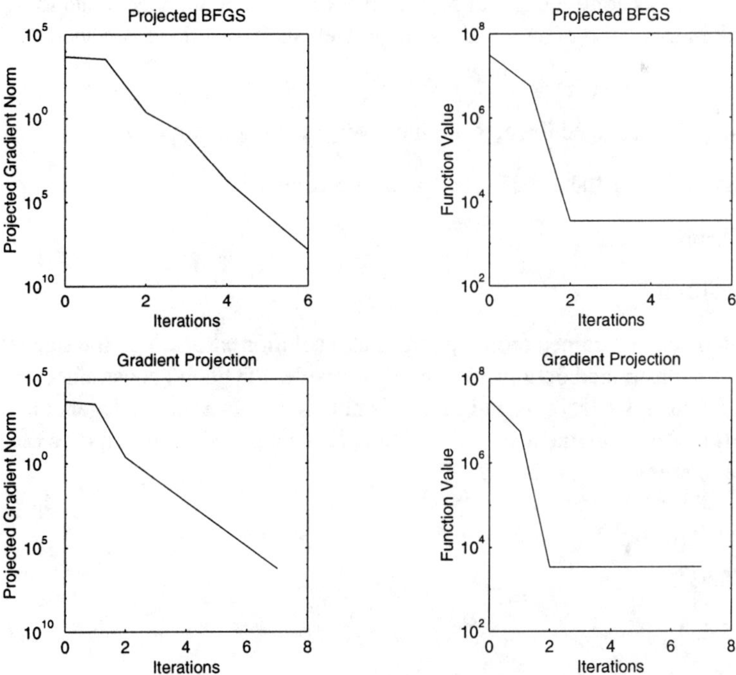

Figure 5.4: *Constrained Discrete Control Problem* II

We solve the constrained problem with Algorithm `gradproj` and Algorithm `bfgsoptb`. In Figure 5.3 we plot the function value and the norm of the projected gradient $u - u(1)$. The projected BFGS iteration required 71 function evaluations, 36 gradient evaluations, and roughly 5.6 million floating point operations, while the gradient projection needed 183 function evaluations, 92 gradient evaluations, and roughly 10.4 million floating point operations.

Our second control problem example solves the same problem as in §3.4.2 using the constraints

$$-206 \leq u \leq 206.$$

We terminate the iteration when $\|u - u(1)\| \leq 10^{-6}$, which is exactly the condition used in §3.4.2 when the active set is empty. The solution to the unconstrained problem is feasible, the active set is empty, and the initial iterate is feasible. Both the gradient projection iteration and the projected BFGS iteration converge to the solution of the unconstrained problem. The constraints are not active at either the initial iterate or the final solution but are active inside the line search for the first iterate and for the second iterate. As is clear from a comparison of Figures 5.4 and 3.3, this small change has a dramatic effect on the cost of the optimization, eliminating the need for the scaling fixup (3.50). The gradient projection method, requiring 15 function evaluations, 8 gradient evaluations, and roughly 167 thousand floating point operations, is far more efficient that the steepest descent iteration reported in §3.4.2. The projected BFGS iteration was somewhat worse, needing 223 thousand operations, but only 13 function evaluations and 7 gradient evaluations. In this example the cost of maintaining the BFGS update was not compensated by a significantly reduced iteration count.

5.8 Exercises on Bound Constrained Optimization

5.8.1. Suppose that f is continuously differentiable, that x^* is a nondegenerate local minimizer for problem (5.4), and *all* constraints are active. Show that there is δ such that

1. if $x \in \mathcal{B}(\delta)$ then $x^* = \mathcal{P}(x - \nabla f(x))$, and
2. the gradient projection algorithm converges in one iteration if $x_0 \in \mathcal{B}(\delta)$.

5.8.2. Show that if $H = I$ then (5.31) and (5.13) are equivalent.

5.8.3. Prove Theorem 5.5.2.

5.8.4. Verify (5.42).

5.8.5. Suppose the unconstrained problem (1.2) has a solution x^* at which the standard assumptions for unconstrained optimization hold. Consider the bound constrained problem (5.3) for u and l such that $x^* \in \Omega$ and $\mathcal{A}(x^*)$ is not empty. Is x^* a nondegenerate local minimizer? If not, how are the results in this chapter changed? You might try a computational example to see what's going on.

5.8.6. Prove Theorem 5.5.4.

5.8.7. Verify (5.51).

5.8.8. Verify (5.52).

5.8.9. Formulate a generalization of (4.33) for updating $A^\dagger$.

5.8.10. What would happen in the examples if we increased the number of (y, s) pairs that were stored? By how much would the BFGS cost be increased?

Part II

Optimization of Noisy Functions

Chapter 6

Basic Concepts and Goals

The algorithms in Part I cannot be implemented at all if the gradient of f is not available, either analytically or via a difference. Even if gradients are available, these algorithms are not satisfactory if f has many local minima that are not of interest. We limit our coverage to deterministic sampling algorithms which are generally applicable and are more or less easy to implement. Of these algorithms, only the DIRECT algorithm [150] covered in §8.4.2 is truly intended to be a global optimizer.

The study of optimization methods that do not require gradients is an active research area (see [227] for a survey of some of this activity), even for smooth problems [61], [62]. Even though some of the methods, such as the Nelder–Mead [204] and Hooke–Jeeves [145] algorithms are classic, most of the convergence analysis in this part of the book was done after 1990.

The algorithms and theoretical results that we present in this part of the book are for objective functions that are perturbations of simple, smooth functions. The surfaces in Figure 6.1 illustrate this problem. The optimization landscape on the left of Figure 6.1, taken from [271], arose in a problem in semiconductor design. The landscape on the right is a simple perturbation of a convex quadratic.

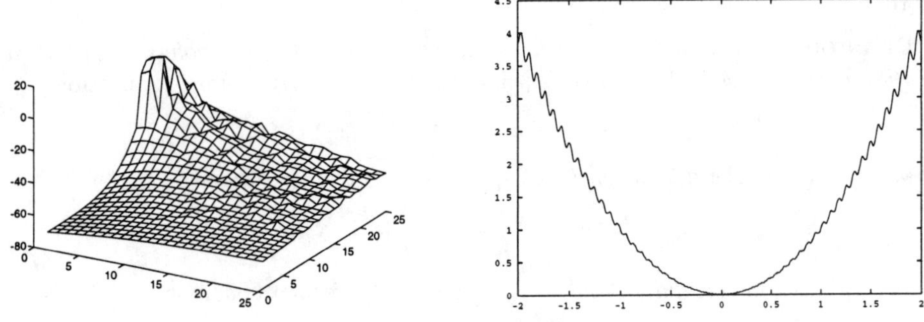

Figure 6.1: *Optimization Landscapes*

We do not discuss algorithms that explicitly smooth the objective function or apply a filter, such as the ones in [168] and [187]. For general problems, these must sample the variable space in some way, for example by performing high-dimensional integration, and are too costly. However, in some special cases these integrals can be performed analytically and impressive results for special-purpose filtering algorithms for computational chemistry have been reported in, for example, [196] and [277]. Nor do we discuss analog methods (see [149] for a well-known example).

We also omit stochastic methods like the special-purpose methods discussed in [38] and [39], or more radical general-purpose global optimization algorithms, such as simulated annealing [166] (see [1] and [265] for surveys of recent work), interval methods [152], or genetic algorithms [143], [144] (see [246] or [123] for a survey), which are random to some extent or random search algorithms. These probabilistic methods, however, should be considered when the more conservative algorithms such as the ones in this part of the book fail.

6.1 Problem Statement

Consider an objective function f that is a perturbation of a smooth function f_s by a small function ϕ

$$f(x) = f_s(x) + \phi(x). \tag{6.1}$$

Small oscillations in ϕ could cause f to have several local minima that would trap any conventional gradient-based algorithms. The perturbation ϕ can, in general, be random or based on the output of an experiment, [250], and may not return the same value when called twice with the same argument. Hence ϕ need not even be a function. We assume that ϕ is everywhere defined and bounded to make the statement of the results simpler.

6.2 The Simplex Gradient

Most of the the algorithms in this part of the book examine a simplex of points in R^N at each iteration and then change the simplex in response. In this section we develop the tools needed to describe and analyze these algorithms. The fundamental idea is that many sampling algorithms require enough information to approximate the gradient by differences and that the accuracy in that difference approximation can be used to analyze the convergence. However, for problems of the form (6.1), one must take care not to make the difference increments so small as to attempt to differentiate the noise.

The ideas in this section were originally used in [155] to analyze the Nelder–Mead [204] algorithm, which we discuss in §8.1. However, the ideas can be applied to several classes of algorithms, and we follow the development in [29] in this section.

DEFINITION 6.2.1. *A* simplex S *in* R^N *is the* convex hull *of* $N+1$ *points,* $\{x_j\}_{j=1}^{N+1}$. x_j *is the* jth vertex *of* S. *We let* V *(or* $V(S)$*) denote the* $N \times N$ *matrix of* simplex directions

$$V(S) = (x_2 - x_1, x_3 - x_1, \ldots, x_{N+1} - x_1) = (v_1, \ldots, v_N).$$

We say S *is* nonsingular *if* V *is nonsingular. The* simplex diameter *diam*(S) *is*

$$diam(S) = \max_{1 \le i,j \le N+1} \|x_i - x_j\|.$$

We will refer to the l^2 *condition number* $\kappa(V)$ *of* V *as the* simplex condition.

We let $\delta(f : S)$ denote the vector of objective function differences

$$\delta(f : S) = (f(x_2) - f(x_1), f(x_3) - f(x_1), \ldots, f(x_{N+1}) - f(x_1))^T.$$

We will not use the simplex diameter directly in our estimates or algorithms. Rather we will use two *oriented lengths*

$$\sigma_+(S) = \max_{2 \le j \le N+1} \|x_1 - x_j\| \text{ and } \sigma_-(S) = \min_{2 \le j \le N+1} \|x_1 - x_j\|.$$

Clearly,

$$\sigma_+(S) \le \text{diam}(S) \le 2\sigma_+(S).$$

6.2.1 Forward Difference Simplex Gradient

DEFINITION 6.2.2. *Let S be a nonsingular simplex with vertices $\{x_j\}_{j=1}^{N+1}$. The* simplex gradient *$D(f : S)$ is*

$$D(f : S) = V^{-T}\delta(f : S).$$

Note that the matrix of simplex directions and the vector of objective function differences depend on which of the vertices is labeled x_1. Most of the algorithms we consider in this part of the book use a vertex ordering or sample on a regular stencil. In this way the algorithms, in one way or another, use a simplex gradient.

This definition of simplex gradient is motivated by the first-order estimate in Lemma 6.2.1.

LEMMA 6.2.1. *Let S be a simplex. Let ∇f be Lipschitz continuous in a neighborhood of S with Lipschitz constant $2K_f$. Then there is $K > 0$, depending only on K_f such that*

$$\|\nabla f(x_1) - D(f : S)\| \le K\kappa(V)\sigma_+(S). \tag{6.2}$$

Proof. Our smoothness assumptions on f and Taylor's theorem imply that for all $2 \le j \le N+1$,

$$|f(x_1) - f(x_j) + v_j^T \nabla f(x_1)| \le K_f\|v_j\|^2 \le K_f\sigma_+(S)^2.$$

Hence

$$\|\delta(f : S) - V^T\nabla f(x_1)\| \le N^{1/2}K_f\sigma_+(S)^2$$

and hence, setting $K = N^{1/2}K_f$,

$$\|\nabla f(x_1) - D(f : S)\| \le K\|V^{-T}\|\sigma_+(S)^2.$$

The conclusion follows from the fact that $\sigma_+(S) \le \|V\|$. □

Search algorithms are not intended, of course, for smooth problems. Minimization of objective functions of the form in (6.1) is one of the applications of these methods. A first-order estimate that takes perturbations into account is our next result.

We will need to measure the perturbations on each simplex. To that end we define for any set T

$$\|\phi\|_T = \sup_{x \in T} \|\phi(x)\|.$$

A first-order estimate also holds for the simplex gradient of an objective function that satisfies (6.1).

LEMMA 6.2.2. *Let S be a nonsingular simplex. Let f satisfy* (6.1) *and let ∇f_s be Lipschitz continuous in a neighborhood of S with Lipschitz constant $2K_s$. Then there is $K > 0$, depending only on K_s, such that*

$$\|\nabla f_s(x_1) - D(f : S)\| \le K\kappa(V)\left(\sigma_+(S) + \frac{\|\phi\|_S}{\sigma_+(S)}\right). \tag{6.3}$$

Proof. Lemma 6.2.1 (applied to f_s) implies

$$\|\nabla f_s(x_1) - D(f_s : S)\| \le K_s N^{1/2}\kappa(V)\sigma_+(S).$$

Now, since $\|\delta(\phi : S)\| \le 2\sqrt{N}\|\phi\|_S$, and $\sigma_+(S) \le \|V\|$,

$$\|D(f : S) - D(f_s : S)\| \le \|V^{-T}\|\|\delta(f : S) - \delta(f_s : S)\| = \|V^{-T}\|\|\delta(\phi : S)\|$$

$$\le 2N^{1/2}\|V^{-T}\|\|\phi\|_S \le 2N^{1/2}\kappa(V)\frac{\|\phi\|_S}{\sigma_+(S)}.$$

This completes the proof with $K = N^{1/2}K_s + 2N^{1/2}$. □

The constants K in (6.2) and (6.3) depend on S only through the Lipschitz constants of f_s and ∇f_s in a neighborhood of S. We will express that dependence as $K = K(S)$ when needed.

The algorithms in this section are most profitably applied to problems of the form (6.1), and the goal is to extract as much information as possible from the smooth part f_s of f without wasting effort in a futile attempt to minimize the noise. In order to formulate our goal for convergence clearly, we explore the consequences of a small simplex gradient in the special (and not uncommon) case that the amplitude of the noise is small in Lemma 6.2.3.

LEMMA 6.2.3. *Let f satisfy* (6.1) *and let ∇f_s be continuously differentiable in a compact set $\Omega \subset R^N$. Assume that f_s has a unique critical point x^* in Ω. Then there is $K_\Omega > 0$ such that for any simplex $S \subset \Omega$ with vertices $\{x_j\}_{j=1}^{N+1}$,*

$$\|x_1 - x^*\| \le K_\Omega \left(\|D(f : S)\| + \kappa(V)\left(\sigma_+(S) + \frac{\|\phi\|_S}{\sigma_+(S)}\right)\right).$$

Proof. The compactness of Ω and our smoothness assumptions on f_s imply that there is β_0 such that

$$\|\nabla f_s(x)\| \ge \beta_0\|x - x^*\|$$

for all $x \in \Omega$. We apply (6.3) to obtain

$$\|x_1 - x^*\| \le \beta_0^{-1}\|\nabla f_s(x_1)\|$$

$$\le \beta_0^{-1}\left(\|D(f : S)\| + K\kappa(V)\left(\sigma_+(S) + \frac{\|\phi\|_S}{\sigma_+(S)}\right)\right).$$

This completes the proof with $K_\Omega = \beta_0^{-1}\max(1, K)$. □

By sampling in an organized way simplex-based algorithms, some directly and some implicitly, attempt to drive the simplex gradient to a small value by changing the size of the simplices over which f is sampled. The motion of the simplices and the scheme for changing the size (especially the reduction in size) accounts for the differences in the algorithms. Theorem 6.2.4, a direct consequence of Lemma 6.2.3, quantifies this. We will consider a sequence of uniformly well-conditioned simplices. Such simplices are generated by several of the algorithms we will study later.

THEOREM 6.2.4. *Let f satisfy* (6.1) *and let ∇f_s be continuously differentiable in a compact set $\Omega \subset R^N$. Assume that f_s has a unique critical point x^* in Ω. Let S^k be a sequence of simplices having vertices $\{x_j^k\}_{j=1}^{N+1}$. Assume that there is M such that*

$$S^k \subset \Omega \text{ and } \kappa(V(S^k)) \le M \text{ for all } k.$$

Then,

1. *if*

$$\lim_{k\to\infty} \sigma_+(S^k) = 0, \lim_{k\to\infty} \frac{\|\phi\|_{S^k}}{\sigma_+(S^k)} = 0,$$

and $\limsup_{k\to\infty} \|D(f : S^k)\| = \epsilon$, *for some* $\epsilon > 0$, *then there is* $K_S > 0$ *such that*

$$\limsup_{k\to\infty} \|x^* - x_1^k\| \le K_S\epsilon;$$

2. *if, for some* $\epsilon > 0$,

$$\limsup_{k\to\infty} \|\phi\|_{S^k} \le \epsilon^2, \liminf_{k\to\infty} \sigma_+(S^k) \ge \epsilon, \text{ and } \liminf_{k\to\infty} \|D(f : S^k)\| \le \epsilon,$$

then there is $K_S > 0$ *such that*

$$\limsup_{k\to\infty} \|x^* - x_1^k\| \le K_S(\epsilon + \limsup_{k\to\infty} \sigma_+(S^k)).$$

6.2.2 Centered Difference Simplex Gradient

In this section we define the centered difference simplex gradient and prove a second-order estimate. We will then prove two variants of Theorem 6.2.4, one to show how the role of the noise ϕ differs from that in the one-sided derivative case and a second to quantify how the values of f on the stencil can be used to terminate an iteration.

DEFINITION 6.2.3. *Let* S *be a nonsingular simplex in* R^N *with vertices* $\{x_j\}_{j=1}^{N+1}$ *and simplex directions* $v_j = x_{j+1} - x_1$. *The* reflected simplex $R = R(S)$ *is the simplex with vertices* x_1 *and*

$$r_j = x_1 - v_j \text{ for } j = 1, \dots, N.$$

The central simplex gradient $D_C(f : S)$ *is*

$$D_C(f : S) = \frac{D(f : S) + D(f : R)}{2} = \frac{V^{-T}(\delta(f : S) - \delta(f : R))}{2}.$$

For example, if $N = 1$ and $x_2 = x_1 + h$, then $r_2 = x_1 - h$. Hence

$$D(f : S) = \frac{f(x_1 + h) - f(x_1)}{h} \text{ and } D(f : R) = \frac{f(x_1 - h) - f(x_1)}{-h}.$$

Therefore,

$$D_C(f : S) = D_C(f : R) = \frac{f(x_1 + h) - f(x_1 - h)}{2h}$$

is the usual central difference.

Lemmas 6.2.5 and 6.2.6 are the second-order analogues of Lemmas 6.2.1 and 6.2.2.

LEMMA 6.2.5. *Let* S *be a nonsingular simplex and let* $\nabla^2 f$ *be Lipschitz continuous in a neighborhood of* $S \cup R(S)$ *with Lipschitz constant* $3K_C$. *Then there is* $K > 0$ *such that*

$$\|\nabla f(x_1) - D_C(f : S)\| \le K\kappa(V)\sigma_+(S)^2. \tag{6.4}$$

Proof. The Lipschitz continuity assumption implies that for all $2 \le j \le N + 1$,

$$|f(x_j) - f(r_j) + 2\nabla f(x_1)^T v_j| \le K_C\|v_j\|^3 \le K_c\sigma_+(S)^3.$$

As in the proof of Lemma 6.2.1 we have

$$\|V^T(\delta(f:S) - \delta(f:R)) - V^T\nabla f(x_1)\| \le N^{1/2} K_C \sigma_+(S)^3,$$

and hence the result follows with $K = N^{1/2}K_C$. □

LEMMA 6.2.6. *Let S be a nonsingular simplex. Let f satisfy* (6.1) *and let $\nabla^2 f_s$ be Lipschitz continuous in a neighborhood of $S \cup R(S)$ with Lipschitz constant $3K_{Cs}$. Then there is $K > 0$, depending only on K_{Cs}, such that*

$$\|\nabla f_s(x_1) - D_C(f:S)\| \le K\kappa(V)\left(\sigma_+(S)^2 + \frac{\|\phi\|_S}{\sigma_+(S)}\right). \tag{6.5}$$

Proof. This proof is very similar to that of Lemma 6.2.2 and is left to the reader. □

The quality of the information that can be obtained from the central simplex gradient is higher than that of the forward. The difference in practice can be dramatic, as the examples in §7.6 illustrate. The consequences of a small central simplex gradient follow directly from Lemma 6.2.6.

LEMMA 6.2.7. *Let f satisfy* (6.1) *and let $\nabla^2 f_s$ be continuously differentiable in a compact set $\Omega \subset R^N$. Assume that f_s has a unique critical point x^* in Ω. Then there is $K_\Omega > 0$ such that if a simplex S and its reflection $R(S)$ are both contained in Ω then*

$$\|x_1 - x^*\| \le K_\Omega\left(\|D_C(f:S)\| + \kappa(V)\left(\sigma_+(S)^2 + \frac{\|\phi\|_S}{\sigma_+(S)}\right)\right).$$

Lemma 6.2.7 is all one needs to conclude convergence from a sequence of small central simplex gradients.

THEOREM 6.2.8. *Let f satisfy* (6.1) *and let $\nabla^2 f_s$ be continuously differentiable in a compact set $\Omega \subset R^N$. Assume that f_s has a unique critical point x^* in Ω. Let S^k be a sequence of simplices having vertices $\{x_j^k\}_{j=1}^{N+1}$. Assume that there is M such that*

$$S^k, R(S^k) \subset \Omega \text{ and } \kappa(V(S^k)) \le M \text{ for all } k.$$

Then,

1. *if*

$$\lim_{k\to\infty} \sigma_+(S^k) = 0, \ \lim_{k\to\infty} \frac{\|\phi\|_{S^k}}{\sigma_+(S^k)} = 0,$$

and $\limsup_{k\to\infty} \|D_C(f:S^k)\| = \epsilon$, *for some* $\epsilon > 0$, *then there is* $K_S > 0$ *such that*

$$\limsup_{k\to\infty} \|x^* - x_1^k\| \le K_S\epsilon;$$

2. *if, for some* $\epsilon > 0$,

$$\limsup_{k\to\infty} \|\phi\|_{S^k} \le \epsilon^3, \liminf_{k\to\infty} \sigma_+(S^k) \ge \epsilon^2, \text{ and } \liminf_{k\to\infty} \|D_C(f:S^k)\| \le \epsilon^2,$$

then there is $K_S > 0$ *such that*

$$\limsup_{k\to\infty} \|x^* - x_1^k\| \le K_S(\epsilon + \limsup_{k\to\infty} \sigma_+(S^k))^2.$$

Theorem 6.2.8, like Theorem 6.2.4, motivates using a small simplex gradient as a test for convergence. Suppose $\|\phi\|_\infty \le \epsilon$ and an algorithm generates sequences of simplices whose vertices are intended to approximate a minimizer of f_s. We can use the results in §2.3.1 to conclude that simplices with $\sigma_+(S) << \epsilon^{1/2}$ will result in inaccurate forward difference gradients and those with $\sigma_+(S) << \epsilon^{2/3}$ in inaccurate central difference gradients. This indicates that the central simplex gradient will be less sensitive to noise than the forward. While this is not usually critical in computing a difference Hessian, where the loss of accuracy may cause slow convergence, it can cause failure of the iteration if one is computing a difference gradient.

If one wants to terminate the algorithm when the simplex gradient is small, say, $\le \tau$, a rough estimate of the minimal possible value of τ is $\tau = O(\epsilon^{1/2})$ for a forward difference simplex gradient and $\tau = O(\epsilon^{2/3})$ for a central simplex gradient.

Moreover, if one is using a centered difference, one has information on the values of f at enough points to make an important qualitative judgment. In order to evaluate a central simplex gradient f must be sampled at x_1 and $x_1 \pm v_j$ for $1 \le j \le N$. If $f(x_1) \le f(x_1 \pm v_j)$ for all $1 \le j \le N$, then one can question the validity of using the simplex gradient as a descent direction or as a measure of stationarity. We call this *stencil failure*. We will use stencil failure as a termination criterion in most of the algorithms we discuss in this part of the book. Our basis for that is a result from [29], which only requires differentiability of f_s.

THEOREM 6.2.9. *Let S be a nonsingular simplex such that for some $\mu_- \in (0,1)$ and $\kappa_+ > 0$,*

$$\kappa(V) \le \kappa_+ \text{ and } x^T V V^T x \ge \mu_- \sigma_+(S)^2 \|x\|^2 \text{ for all } x. \tag{6.6}$$

Let f satisfy (6.1) *and let ∇f_s be Lipschitz continuously differentiable in a ball B of radius $2\sigma_+(S)$ about x_1. Assume that*

$$f(x_1) < \min_j \{f(x_1 \pm v_j)\}. \tag{6.7}$$

Then, if K is the constant from Lemma 6.2.2,

$$\|\nabla f_s(x_1)\| \le 8\mu_-^{-1} K \kappa_+ \left(\sigma_+(S) + \frac{\|\phi\|_B}{\sigma_+(S)}\right). \tag{6.8}$$

Proof. Let $R(S)$, the reflected simplex, have vertices x_1 and $\{r_j\}_{j=1}^N$. (6.7) implies that each component of $\delta(f:S)$ and $\delta(f:R)$ is positive. Now since

$$V = V(S) = -V(R),$$

we must have

$$\begin{aligned} 0 &< \delta(f:S)^T \delta(f:R) \\ &= (V^T V^{-T} \delta(f:S))^T (V(R)^T V(R)^{-T} \delta(f:R)) \\ &= -D(f:S)^T V V^T D(f:R). \end{aligned} \tag{6.9}$$

We apply Lemma 6.2.2 to both $D(f:S)$ and $D(f:R)$ to obtain

$$D(f:S) = \nabla f_s(x_1) + E_1 \text{ and } D(f:R) = \nabla f_s(x_1) + E_2,$$

where, since $\kappa(V) = \kappa(V(R)) \le \kappa_+$,

$$\|E_k\| \le K\kappa_+ \left(\sigma_+(S) + \frac{\|\phi\|_B}{\sigma_+(S)}\right).$$

Since $\|V\| \le 2\sigma_+(S)$ we have by (6.9)

$$\begin{aligned} \nabla f_s(x_1)^T V V^T \nabla f_s(x_1) \le\ & 4\sigma_+(S)^2 \|\nabla f_s(x_1)\| (\|E_1\| + \|E_2\|) \\ & + 4\sigma_+(S)^2 \|E_1\| \|E_2\|. \end{aligned} \tag{6.10}$$

The assumptions of the lemma give a lower estimate of the left side of (6.10),

$$w^T V V^T w \ge \mu_- \sigma_+(S)^2 \|w\|^2.$$

Hence,

$$\|\nabla^2 f(x_1)\| \le b\|\nabla^2 f(x_1)\| + c,$$

where, using (6.10),

$$b = 8\mu_1^{-1} K_s \kappa_+ \left(\sigma_+(S) + \frac{\|\phi\|_B}{\sigma_+(S)} \right)$$

and

$$c = 4\mu_-^{-1} (K_s \kappa_+)^2 \left(\sigma_+(S) + \frac{\|\phi\|_B}{\sigma_+(S)} \right)^2 = \frac{\mu_-}{16} B^2.$$

So $b^2 - 4c = b^2(1 - \mu_-/4)$ and the quadratic formula then implies that

$$\|\nabla^2 f(x_1)\| \le \frac{b + \sqrt{b^2 - 4c}}{2} = b\frac{1 + \sqrt{1 - \mu_-/4}}{2} \le b$$

as asserted. □

6.3 Examples

Our examples are selected to represent a variety of problems that can be attacked by the methods in this part of the book and, at the same time, are easy for the reader to implement. Many of the problems to which these methods have been applied have complex objective functions and have been solved as team efforts [107], [250], [121], [70], [69]. In many such cases the objective function is not even available as a single subroutine as the optimizer, simulator, and design tool are one package. Hence, the examples we present in this part of the book are even more artificial than the ones in the first part. The cost of an evaluation of f is much less in these examples than it is in practice.

6.3.1 Weber's Problem

Our discussion of this problem is based on [182]. *Weber's problem* is to locate a central facility (a warehouse or factory, for example) so that the total cost associated with distribution to several demand centers is minimized. The model is that the cost is proportional to the distance from the facility. The proportionality constant may be positive reflecting transportation costs or negative reflecting environmental concerns.

If the locations of the demand centers are $\{z_i\} \subset R^2$ and the corresponding weights are $\{w_i\}$, then the objective function is

$$f(x) = \sum_i w_i \|x - z_i\| = \sum_i w_i \sqrt{[(x)_1 - (z_i)_1]^2 + [(x)_2 - (z_i)_2]^2}. \tag{6.11}$$

We will assume that

$$\sum_i w_i > 0,$$

so that a global optimum exists. If $\sum_i w_i < 0$ then $\inf f = -\infty$ and there is no global optimum.

Weber's problem is not differentiable at $x = z_i$ because the square root function is not differentiable at 0. A gradient-based algorithm, applied in a naive way, will have difficulty with this problem. There are special-purpose algorithms (see [182] for a survey) for Weber's problem, especially if all the weights are positive. Our main interest is in the case where at least one weight is negative. In that case there may be multiple local minima.

We will consider two examples. The first, and simplest, is from [182]. This example has three demand centers with

$$w = (2, 4, -5)^T \text{ and } (z_1, z_2, z_3) = \begin{pmatrix} 2 & 90 & 43 \\ 42 & 11 & 88 \end{pmatrix}.$$

The global minimum is at $x^* = (90, 11)^T$, at which the gradient is not defined. The complex contours near the minimizer in Figure 6.2 illustrate the difficulty of the problem.

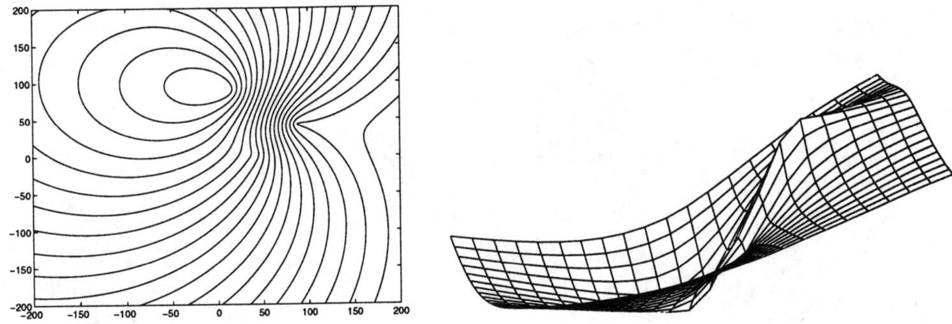

Figure 6.2: *Contour/Surface for Weber's Function: First Example*

Our second example has two local minimizers, at $(-10, -10)$ and $(25, 30)$ with the global minimizer at $(25, 30)$. There are four demand centers with

$$w = (2, -4, 2, 1)^T \text{ and } (z_1, z_2, z_3, z_4) = \begin{pmatrix} -10 & 0 & 5 & 25 \\ -10 & 0 & 8 & 30 \end{pmatrix}.$$

See Figure 6.3.

Our third example adds the oscillatory function

$$\phi(x) = \sin(.0035x^T x) + 5\sin(.003(x - y)^T (x - y))$$

to the second example, where $y = (-20, 0)^T$. This complicates the optimization landscape significantly, as the surface and contour plots in Figure 6.4 show.

6.3.2 Perturbed Convex Quadratics

The sum of a simple convex quadratic and low-amplitude high-frequency perturbation will serve as a model problem for all the algorithms in this section. For example, the function graphed on the right in Figure 6.1,

$$f(x) = 2x^2(1 + .75\cos(80x)/12) + \cos(100x)^2/24$$

is one of the examples in [120]. Our general form will be

$$(6.12)\quad \begin{aligned} f(x) &= (x - \xi_0)^T H(x - \xi_0)(1 + a_1\cos(b_1^T(x - \xi_1) + c_1(x - \xi_1)^T(x - \xi_1))) \\ &\quad + a_2(1 + \cos(b_2^T(x - \xi_2)^T + c_2(x - \xi_2)^T(x - \xi_2))) + a_3|\text{rand}|, \end{aligned}$$

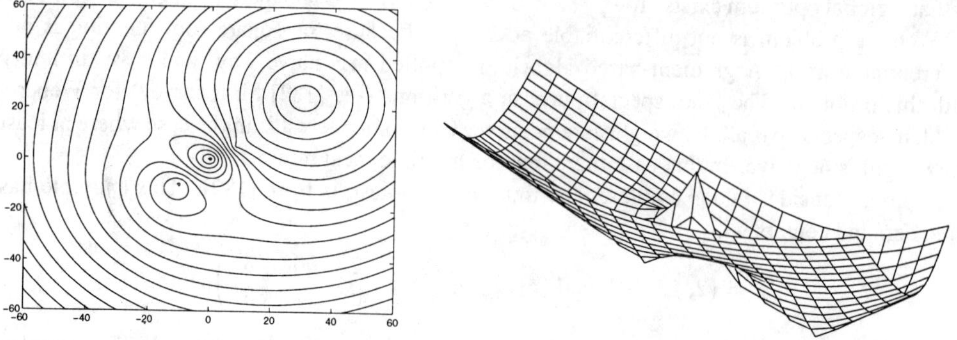

Figure 6.3: *Contour/Surface for Weber's Function: Second Example*

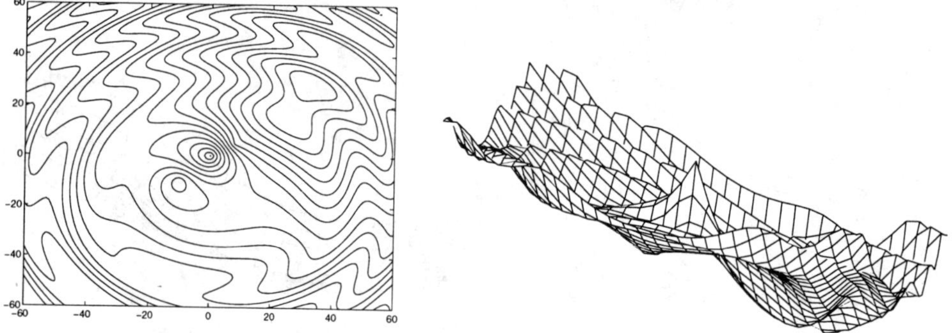

Figure 6.4: *Contour/Surface plots for Weber's Function: Third Example*

where $\{\xi_j\}$, $\{a_j\}$, $\{b_j\}$, $\{c_j\}$ are given and rand is a random number generator. f has been designed so that the minimum value is $O(a_1+a_2+a_3)$. The unperturbed case $a_1 = a_2 = a_3 = 0$ is also of interest for many of the algorithms in this part of the book.

6.3.3 Lennard–Jones Problem

The objective function is a simple model of the potential energy in a molecule of identical atoms. Assume that there are M atoms and that $\xi_i \in R^3$ is the position of the ith atom. Letting

$$d_{ij} = \|\xi_i - \xi_j\|$$

and

$$x = (\xi_1^T, \ldots, \xi_M^T)^T \in R^N$$

where $N = 3M$, we have that the Lennard–Jones energy function is

$$f(x) = \sum_{i=1}^{M} \sum_{j=1}^{i-1} \left(d_{ij}^{-12} - 2d_{ij}^{-6} \right). \tag{6.13}$$

f has many local minimizers ($O(e^{M^2})$ is one conjecture [142]) and the values at the minimizers are close. Hence, the Lennard–Jones function does not conform to the noisy perturbation of a smooth function paradigm. The reader is asked in some of the exercises to see how the methods perform.

6.4 Exercises on Basic Concepts

6.4.1. Show that if $w_i > 0$ for all i then Weber's problem has a unique local minimum.

6.4.2. Prove Lemma 6.2.6.

6.4.3. Try to minimize the Lennard–Jones functional using some of the algorithms from the first part of the book. Vary the initial iterate and M. Compare your best results with those in [142], [40], and [210].

Chapter 7

Implicit Filtering

7.1 Description and Analysis of Implicit Filtering

The implicit filtering algorithm was originally formulated in [270], [251], and [271], as a difference-gradient implementation of the gradient projection algorithm [18] in which the difference increment is reduced in size as the iteration progresses. A different formulation for unconstrained problems with certain convexity properties was introduced at about the same time in [279]. From the point of view of this book, the simplex gradient is used in a direct way. The algorithmic description and analysis in this chapter uses the results from §6.2 directly. We will focus on unconstrained problems and derive the convergence results that implicit filtering shares with the search algorithms in Chapter 8.

Implicit filtering, by using an approximate gradient directly, offers the possibility of improved performance with quasi-Newton methods and can be easily applied to bound constrained problems. We explore these two possibilities in §§7.2 and 7.4.

In its simplest unconstrained form, implicit filtering is the steepest descent algorithm with difference gradients, where the difference increment varies as the iteration progresses. Because the gradient is only an approximation, the computed steepest descent direction may fail to be a descent direction and the line search may fail. In this event, the difference increment is reduced.

For a given $x \in R^N$ and $h > 0$ we let the simplex $S(x, h)$ be the right simplex from x with edges having length h. Hence the vertices are x and $x + hv_i$ for $1 \le i \le N$ with $V = I$. So $\kappa(V) = 1$. The performance of implicit filtering with a central difference gradient is far superior to that with the forward difference gradient [120], [187], [250]. We will, therefore, use centered differences in the discussion. We illustrate the performance of forward difference gradients in §7.6.

We set

$$\nabla_h f(x) = D_C(f : S(x, h)).$$

We use a simple Armijo [7] line search and demand that the sufficient decrease condition

$$f(x - \lambda \nabla_h f(x)) - f(x) < -\alpha\lambda \|\nabla_h f(x)\|^2 \tag{7.1}$$

holds (compare with (3.4)) for some $\alpha > 0$.

Our central difference steepest descent algorithm `fdsteep` terminates when

$$\|\nabla_h f(x)\| \le \tau h \tag{7.2}$$

for some $\tau > 0$, when more than $pmax$ iterations have been taken, after a stencil failure, or when the line search fails by taking more than $amax$ backtracks. Even the failures of `fdsteep` can be used to advantage by triggering a reduction in h. The line search parameters α, β and the parameter τ in the termination criterion (7.2) do not affect the convergence analysis that we present here but can affect performance.

ALGORITHM 7.1.1. fdsteep($x, f, pmax, \tau, h, amax$)

1. *For* $p = 1, \dots, pmax$

   (a) *Compute* f *and* $\nabla_h f$; *terminate if* (6.7) *or* (7.2) *hold.*

   (b) *Find the least integer* $0 \le m \le amax$ *such that* (7.1) *holds for* $\lambda = \beta^m$. *If no such* m *exists, terminate.*

   (c) $x = x - \lambda \nabla_h f(x)$.

Algorithm fdsteep will terminate after finitely many iterations because of the limits on the number of iterations and the number of backtracks. If the set $\{x \mid f(x) \le f(x_0)\}$ is bounded then the iterations will remain in that set. Implicit filtering calls fdsteep repeatedly, reducing h after each termination of fdsteep. Aside from the data needed by fdsteep, one must provide a sequence of difference increments, called *scales* in [120].

ALGORITHM 7.1.2. imfilter1($x, f, pmax, \tau, \{h_k\}, amax$)

1. *For* $k = 0, \dots$
   Call fdsteep($x, f, pmax, \tau, h_k, amax$)

The convergence result follows from the second-order estimate, (6.5), the consequences of a stencil failure, Theorem 6.2.9, and the equalities $h_k = \sigma_+(S^k)$ and $\kappa(V^k) = 1$. A similar result for forward differences would follow from (6.3).

THEOREM 7.1.1. *Let* f *satisfy* (6.1) *and let* ∇f_s *be Lipschitz continuous. Let* $h_k \to 0$, $\{x_k\}$ *be the implicit filtering sequence, and* $S^k = S(x, h_k)$. *Assume that* (7.1) *holds (i.e., there is no line search failure) for all but finitely many* k. *Then if*

$$\lim_{k \to \infty} (h_k + h_k^{-1} \|\phi\|_{S^k}) = 0 \tag{7.3}$$

then any limit point of the sequence $\{x_k\}$ *is a critical point of* f_s.

Proof. If either (7.1) or (6.7) hold for all but finitely many k then, as is standard,

$$\nabla_{h_k} f(x_k) = D_C(f : S^k) \to 0.$$

Hence, using (7.3) and Lemma 6.2.2,

$$\nabla f_s(x_k) \to 0,$$

as asserted. □

7.2 Quasi-Newton Methods and Implicit Filtering

The unique feature of implicit filtering is the possibility, for problems that are sufficiently smooth near a minimizer, to obtain faster convergence in the terminal phase of the iteration by using a quasi-Newton update of a model Hessian. This idea was first proposed in [250] and [120].

We begin with a quasi-Newton form of Algorithm fdsteep. In this algorithm a quasi-Newton approximation to the Hessian is maintained and the line search is based on the quasi-Newton direction

$$d = -H^{-1} \nabla_h f(x)$$

terminating when either

$$f(x + \lambda d) - f(x) < \alpha\lambda\nabla_h f(x)^T d \tag{7.4}$$

or too many stepsize reductions have been taken. With the application to implicit filtering in mind, Algorithm fdquasi replaces the quasi-Newton H with the identity matrix when the line search fails.

ALGORITHM 7.2.1. fdquasi$(x, f, H, pmax, \tau, h, amax)$

1. *For* $p = 1, \ldots, pmax$
   (a) *Compute* f, $\nabla_h f$ *and* $d = -H^{-1}\nabla_h f$*; terminate if* (7.2) *holds.*
   (b) *Find the least integer* $0 \le m \le amax$ *such that* (7.4) *holds for* $\lambda = \beta^m$.
   (c) $x = x + \lambda d$.
   (d) *Update* H *with a quasi-Newton formula.*

In the context of implicit filtering, where N is small, the full quasi-Newton Hessian or its inverse is maintained throughout the iteration. Our MATLAB codes store the model Hessian.

ALGORITHM 7.2.2. imfilter2$(x, f, pmax, \tau, \{h_k\}, amax)$

1. $H = I$.
2. *For* $k = 0, \ldots$
   Call fdquasi$(x, f, H, pmax, \tau, h_k, amax)$.

In [250] and [120] the SR1 method was used because it performed somewhat better than the BFGS method in the context of a particular application. The examples in §7.6 show the opposite effect, and both methods have been successfully used in practice.

7.3 Implementation Considerations

Implicit filtering has several iterative parameters and requires some algorithmic decisions in its implementation. The parameters $pmax$, $amax$, and β play the same role that they do in any line search algorithm. In our MATLAB code imfil.m, which we used for all the computations reported in this book, we set $pmax = 200 * n$, $amax = 10$, and $\beta = 1/2$.

The performance of implicit filtering can be sensitive to the value of τ [250], with small values of τ leading to stagnation and values of τ that are too large leading to premature termination of fdquasi. Using stencil failure as a termination criterion reduces the sensitivity to small values of τ and we use $\tau = .01$ in the computations.

The sequence of scales is at best a guess at the level of the noise in the problem. If several of the scales are smaller than the level of the noise, the line search will fail immediately and work at these scales will be wasted. Our implementation attempts to detect this by terminating the optimization if the x is unchanged for three consecutive scales.

The simplex gradient may be a very poor approximation to the gradient. In some such cases the function evaluation at a trial point may fail to return a value [250] and one must either trap this failure and return an artificially large value, impose bound constraints, or impose a limit on the size of the step. In our computations we take the latter approach and limit the stepsize to $10h$ by setting

$$d = \begin{cases} -H^{-1}\nabla_h f(x) & \text{if } \|H^{-1}\nabla_h f(x)\| \le 10h, \\ \dfrac{-10hH^{-1}\nabla_h f(x)}{\|H^{-1}\nabla_h f(x)\|} & \text{otherwise.} \end{cases} \tag{7.5}$$

The choice of a quasi-Newton method to use with implicit filtering is an area of active research [56], [55]. Both SR1 and BFGS have been used, with SR1 performing modestly better in some applications with bound constraints [270], [251], [271], [250], [55]. The implementation of implicit filtering in the collection of MATLAB codes `imfil.m` uses BFGS as the default but has SR1 as an option. We found BFGS with central differences to be consistently better in the preparation of the (unconstrained!) computational examples in this book.

7.4 Implicit Filtering for Bound Constrained Problems

Implicit filtering was initially designed as an algorithm for bound constrained problems [250], [120]. The bound constrained version we present here is simply a projected quasi-Newton algorithm like the one presented in §5.5.3. There are other approaches to the implementation and no best approach has emerged. We refer the reader to [120] and [55] for discussions of the options.

We begin with scaling and the difference gradient. Central differences perform better, but we do not evaluate f outside of the feasible region. Hence, if a point on the centered difference stencil is outside of the feasible region, we use a one-sided difference in that direction. In order to guarantee that at least one point in each direction is feasible, we scale the variables so that $L_i = 0$, $U_i = 1$, and $h_0 \leq 1/2$.

The sufficient decrease condition is (compare with (5.31))

$$f(x(\lambda)) - f(x) \leq \alpha \nabla_h f(x)^T (x(\lambda) - x), \tag{7.6}$$

where

$$x(\lambda) = \mathcal{P}(x - \lambda \nabla_h f(x)).$$

One could terminate the iteration at a given scale when the analogue to (7.2)

$$\|x - x(1)\| \leq \tau h \tag{7.7}$$

holds or when

$$f(x_c) < f(x \pm r_j) \text{ for all } x \pm r_j \text{ feasible}, \tag{7.8}$$

which is the analogue to (6.7) for bound constrained problems.

Quasi-Newton methods for bound constraints can be constructed more simply for small problems, like the ones to which implicit filtering is applied, where it is practical to store the model of the inverse of the reduced Hessian as a full matrix. By using full matrix storage, the complexity of `bfgsrecb` is avoided. One such alternative [53], [54], [55] to the updates in §5.5.3 is to update the complete reduced Hessian and then correct it with information from the new active set. This results in a two-stage update in which a model for the inverse of reduced Hessian is updated with (4.5) to obtain

$$R_{1/2}^{-1} = \left(I - \frac{sy^T}{y^T s}\right) R_c^{-1} \left(I - \frac{ys^T}{y^T s}\right) + \frac{ss^T}{y^T s}. \tag{7.9}$$

Then the new reduced Hessian is computed using the active set information at the new point

$$R_+^{-1} = \mathcal{P}_{\mathcal{A}_+} + \mathcal{P}_{\mathcal{I}_+} R_{1/2}^{-1} \mathcal{P}_{\mathcal{I}_+}. \tag{7.10}$$

It is easy to show that Theorems 5.5.4 and 5.5.5 hold for this form of the update.

A FORTRAN implementation [119] of implicit filtering for bound constrained problems is in the software collection. In the original version of that implementation a projected SR1 update was used and a Cholesky factorization of the matrix R_+ was performed to verify positivity. The model Hessian was reinitialized to the identity whenever the scale or the active set changed.

7.5 Restarting and Minima at All Scales

No algorithm in this part of the book is guaranteed to find even a local minimum, much less a global one. One approach to improving the robustness of these algorithms is to restart the iteration after one sweep through the scales. A point x that is not changed after a call to Algorithm `imfilter1` (or `imfilter2` or the bound constrained form of either) is called a *minimum at all scales*.

If f satisfies (6.1), f_s has a unique critical point that is also a local minimum that satisfies the standard assumptions (and hence is a global minimum for f_s), and certain (strong!) technical assumptions on the decay of ϕ near the minimum hold, then [120] a minimum at all scales is near that global minimum of f_s. In the unconstrained case this statement follows from the termination criteria ((7.2) and (6.7)) for implicit filtering, Lemma 6.2.3 (or 6.2.7) and, if central differences are used, Theorem 6.2.9. The analysis in [120] of the bound constrained case is more technical.

In practice, restarts are expensive and need not be done for most problems. However, restarts have been reported to make a difference in some cases [178]. It is also comforting to know that one has a minimum at all scales, and the author of this book recommends testing potential optima with restarts before one uses the results in practice but not at the state where one is tuning the optimizer or doing preliminary evaluation of the results.

7.6 Examples

Many of these examples are from [56]. For all the examples we report results with and without a quasi-Newton Hessian. We report results for both forward and central differences. In the figures the solid line corresponds to the BFGS Hessian, the dashed-dotted line to the SR1 Hessian, and the dashed line to $H = I$, the steepest descent form of implicit filtering.

Unlike the smooth problems considered earlier, where convergence of the gradient to zero was supported by theory, convergence of the simplex gradient to zero is limited by the noise in the objective. We illustrate performance by plotting both the objective function value and the norm of the simplex gradient. From these examples it is clear that the the graphs of function value against the count of function evaluations is a better indicator of the performance of the optimizer.

In all cases we terminated the iteration when either `fdquasi` had been called for each scale or a budget of function evaluations had been exhausted. Once the code completes an iteration and the number of function evaluations is greater than or equal to the budget, the iteration is terminated.

The examples include both smooth and nonsmooth problems, with and without noise. A serious problem for some algorithms of this type is their failure on very easy problems. For most of the algorithms covered in this part of the book, we will present examples that illustrate performance on this collection of problems.

7.6.1 Weber's Problem

The three Weber's function examples all have minimizers at points at which the objective is nondifferentiable. For the computations we used an initial iterate of $(10, -10)^T$, a budget of 200 function evaluations, and $\{10 \times 2^{-n}\}_{n=-2}^{8}$ as the sequence of scales.

In each of the examples the performance of the two quasi-Newton methods was virtually identical and far better than that without a quasi-Newton model Hessian. Forward and central differences for the first two problems (Figures 7.1 and 7.2) perform almost equally well, with forward having a slight edge. In Figure 7.3, however, the forward difference version of implicit filtering finds a local minimum different from the global minimum that is located by central

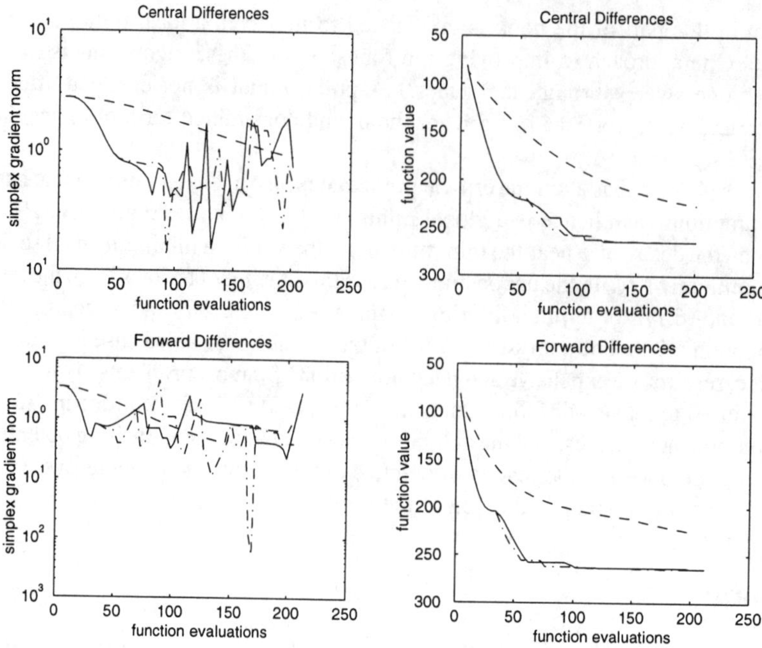

Figure 7.1: *First Weber Example*

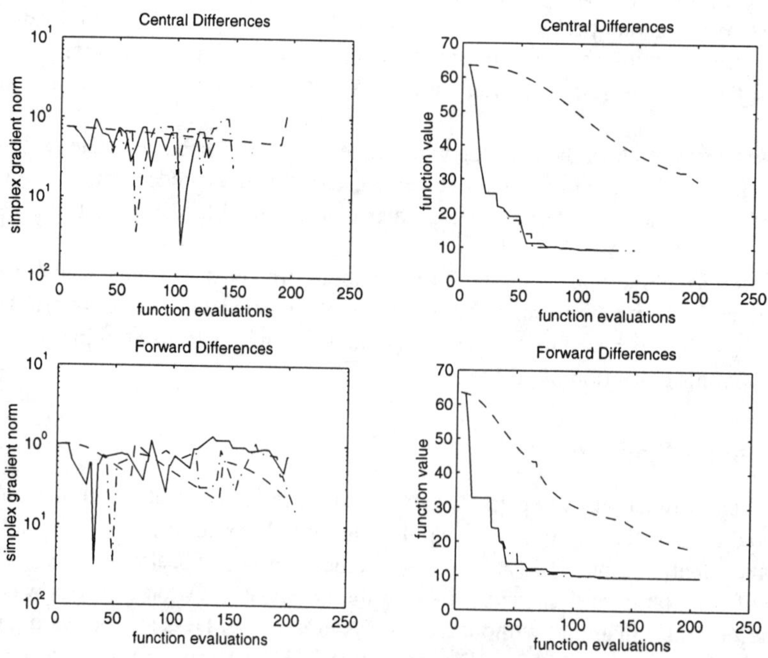

Figure 7.2: *Second Weber Example*

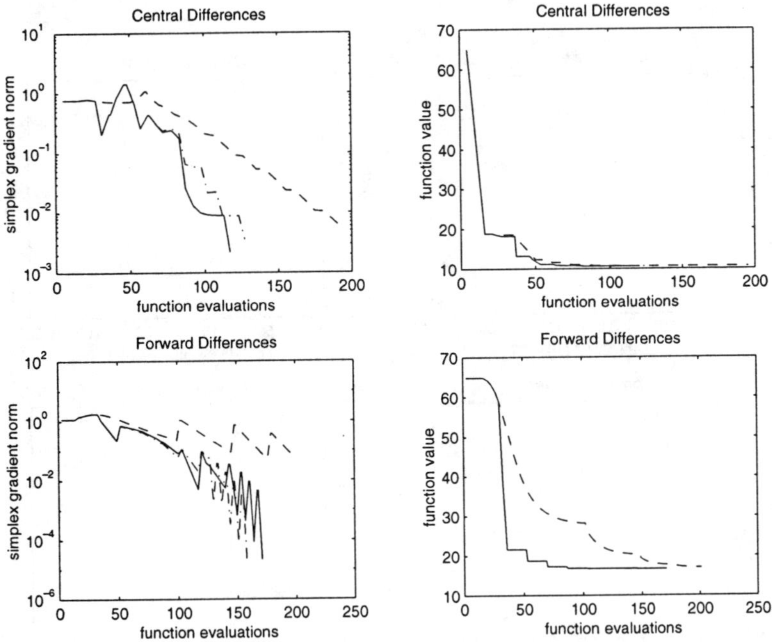

Figure 7.3: *Third Weber Example*

differencing. This, of course, is consistent with the theory, which does not claim that implicit filtering is a global optimizer.

7.6.2 Parameter ID

We consider the parameter ID example from §1.6.2 using the data from §2.6.1. Recall that in this example we use as data the values of the exact solution for $c = k = 1$ at the points $t_i = i/100$ for $1 \le i \le 100$. The initial iterate was $(5,5)^T$; the sequence of scales was $\{2^{-k}\}_{k=1}^{12}$. Implicit filtering, like the globally convergent algorithms in the first part of the book, is fairly insensitive to the choice of initial iterate, as we will see when we revisit this example in §8.5.2.

We report on both low ($rtol = atol = 10^{-3}$, Figure 7.4) and high ($rtol = atol = 10^{-6}$, Figure 7.5) accuracy computations. Note that after 200 function evaluations the function reduction from the central difference BFGS form of implicit filtering flattens out in both plots at roughly the expected level of $O(tol)$ while the other methods have not. This effect, which is not uncommon, is one reason for our preference for the BFGS central difference form of the algorithm.

7.6.3 Convex Quadratics

The performance of the central difference BFGS form of implicit filtering should be very good, since (see exercises 7.7.1 and 7.7.2) the difference approximation of the gradient is exact. We would expect that good performance to persist in the perturbed case. We illustrate this with results on two problems, both given by (6.12). One is an unperturbed problem ($a_j = b_j = c_j = 0$ for all j) where H is a diagonal matrix with $(H)_{ii} = 1/(2i)$ for $1 \le i \le N$. The other is a perturbed problem with

$$\xi_0 = (\sin(1), \sin(2), \ldots, \sin(N))^T, \xi_1 = 0, \xi_2 = (1, \ldots, 1)^T,$$

$$a_1 = a_2 = .01, a_3 = 0, b_1 = (1, \ldots, 1)^T, b_2 = 0, \text{ and } c_1 = c_2 = 10\pi.$$

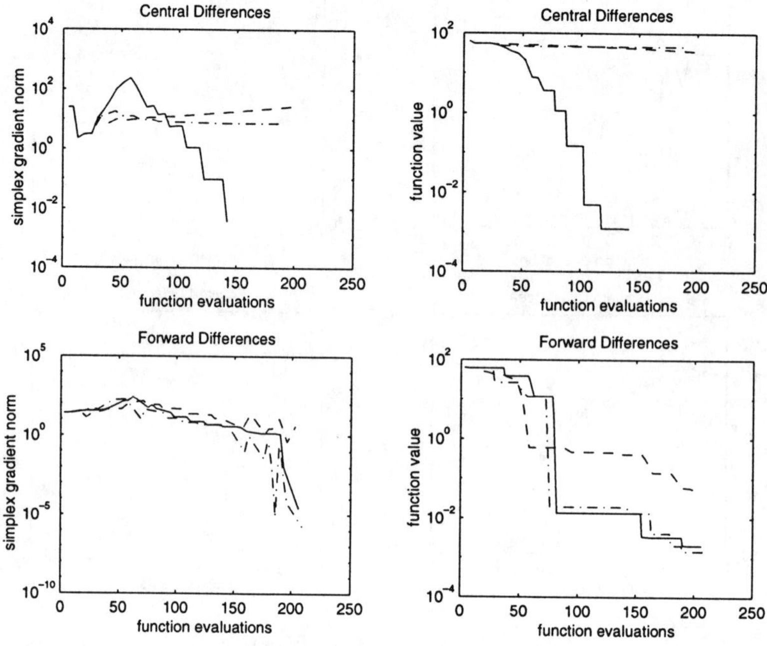

Figure 7.4: *Parameter ID, tol* $= 10^{-3}$

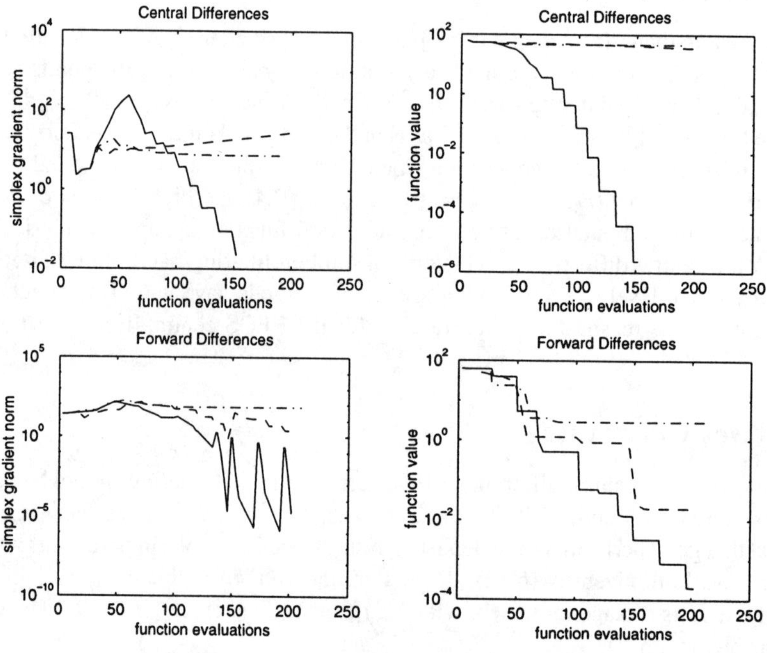

Figure 7.5: *Parameter ID, tol* $= 10^{-6}$

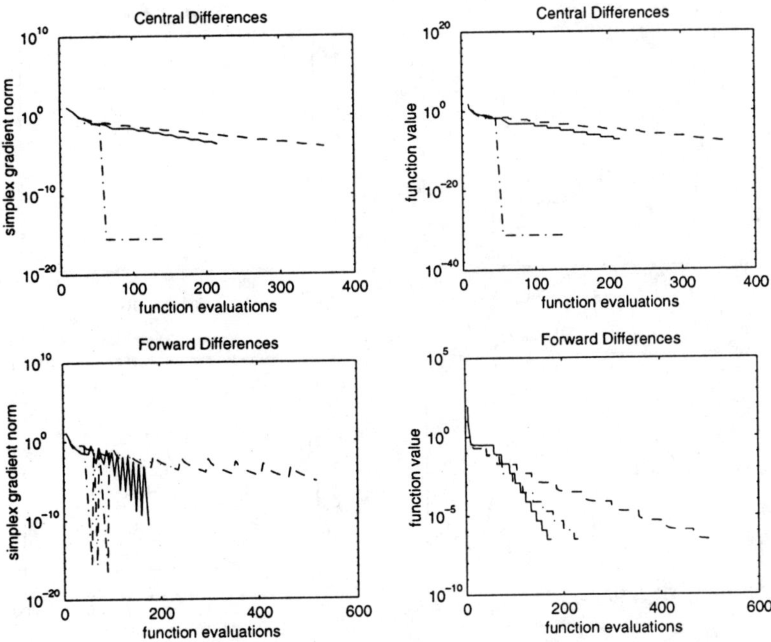

Figure 7.6: *Unperturbed Quadratic,* $N = 4$

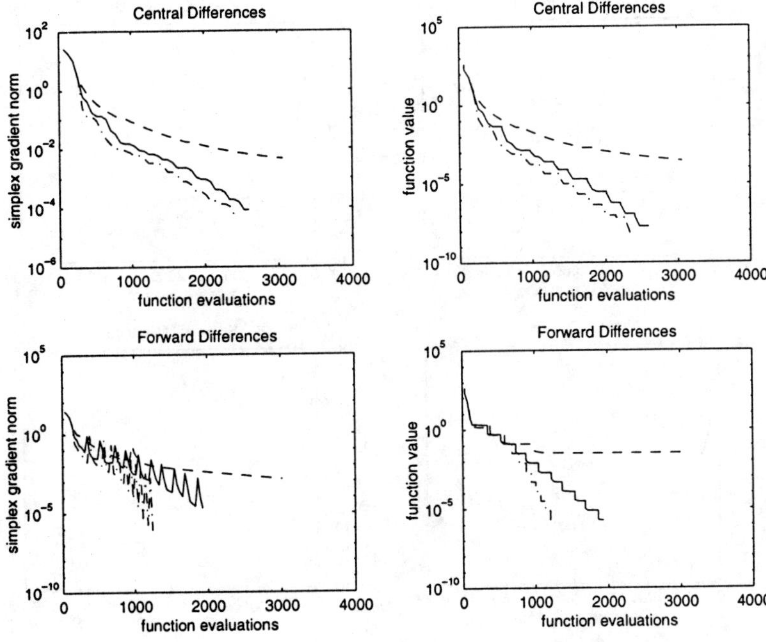

Figure 7.7: *Unperturbed Quadratic,* $N = 32$

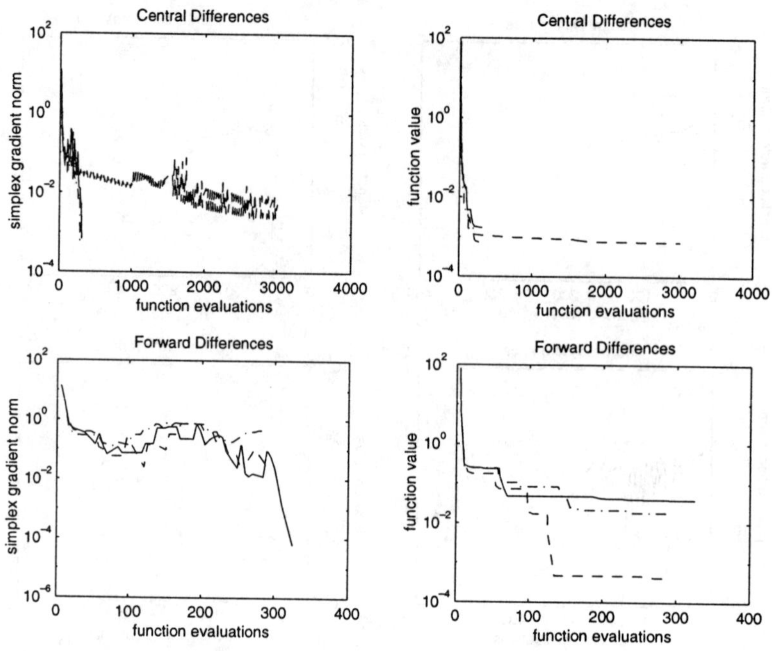

Figure 7.8: *Perturbed Quadratic,* $N = 4$

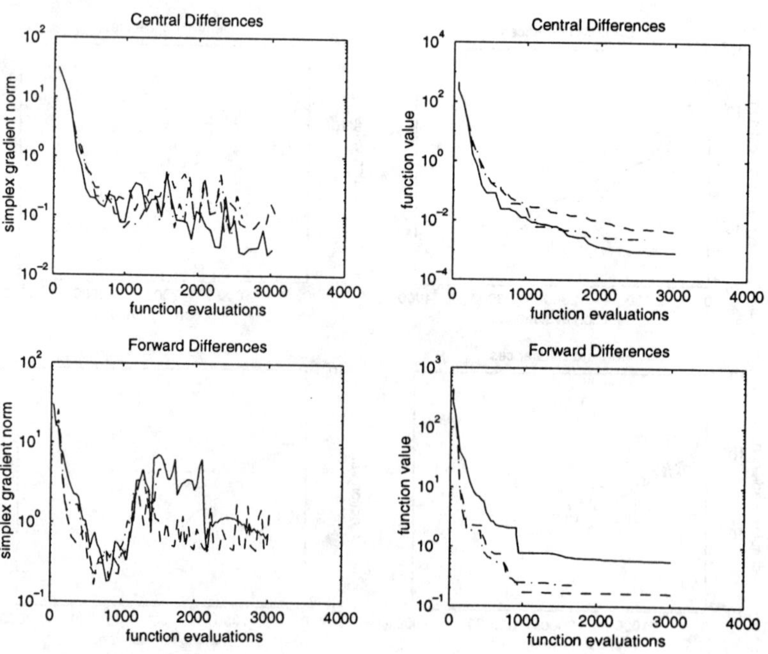

Figure 7.9: *Perturbed Quadratic,* $N = 32$

If $a_3 \neq 0$ then f may not return the same value when called with the same argument twice. The reader is invited to explore the consequences of this in exercise 7.7.3.

The performance of the algorithms in this part of the book sometimes depends on the size of the problem much more strongly than the Newton-based methods in Part I. In the case of implicit filtering, that dependence is mostly a result of the cost of evaluation of the simplex gradient. To illustrate this we consider our quadratic problems for $N = 4$ (Figures 7.6 and 7.8) and $N = 32$ (Figures 7.7 and 7.9).

For all the quadratic examples the initial iterate was

$$x_0 = \frac{(1, 2, \ldots, N)^T}{10N}$$

and the sequence of scales was $\{2^{-k}\}_{k=0}^{10}$.

7.7 Exercises on Implicit Filtering

7.7.1. Let S be a nonsingular simplex. Show that $D_C(f : S) = f(x_1)$ if f is a quadratic function.

7.7.2. How would you expect forward and centered difference implicit filtering to perform when applied to $f(x) = x^T x$? Would the performance be independent of dimension? Test your expectation with numerical experimentation.

7.7.3. Use implicit filtering to minimize the perturbed quadratic function with nonzero values of a_3.

7.7.4. Try to solve the Lennard–Jones problem with implicit filtering for various values of M and various initial iterates. Compare your best results with those in [142], [40], and [210]. Are you doing any better than you did in exercise 6.4.3?

7.7.5. Show that Theorems 5.5.4 and 5.5.5 hold if the projected BFGS update is implemented using (7.9) and (7.10). How would these formulas affect an implementation like `bfgsrecb`, which is designed for problems in which full matrices cannot be stored?

Chapter 8

Direct Search Algorithms

In this chapter we discuss the class of *direct search algorithms*. These methods use values of f taken from a set of sample points and use that information to continue the sampling. Unlike implicit filtering, these methods do not explicitly use approximate gradient information. We will focus on three such methods: the *Nelder–Mead simplex algorithm* [204], the *multidirectional search method* [85], [261], [262], and the *Hooke–Jeeves algorithm* [145]. Each of these can be analyzed using the simplex gradient techniques from Chapter 6. We will not discuss the very general results based on the taxonomies of direct search methods from [263], [174], and [179] or the recent research on the application of these methods to bound [173] or linear [175] constraints.

We include at the end of this chapter a short discussion of methods based on surrogate models and a brief account of a very different search method, the *DIRECT algorithm* [150]. These two final topics do not lead to algorithms that are easy to implement, and our discussions will be very general with pointers to the literature.

8.1 The Nelder–Mead Algorithm

8.1.1 Description and Implementation

The Nelder–Mead [204] simplex algorithm maintains a simplex S of approximations to an optimal point. In this algorithm the vertices $\{x_j\}_{j=1}^{N+1}$ are sorted according to the objective function values

$$f(x_1) \leq f(x_2) \leq \cdots \leq f(x_{N+1}). \tag{8.1}$$

x_1 is called the best vertex and x_{N+1} the worst. If several vertices have the same objective value as x_1, the best vertex is not uniquely defined, but this ambiguity has little effect on the performance of the algorithm.

The algorithm attempts to replace the worst vertex x_{N+1} with a new point of the form

$$x(\mu) = (1 + \mu)\overline{x} - \mu x_{N+1}, \tag{8.2}$$

where $\overline{x}$ is the centroid of the convex hull of $\{x_i\}_{i=1}^{N}$

$$\overline{x} = \frac{1}{N}\sum_{i=1}^{N} x_i. \tag{8.3}$$

The value of μ is selected from a sequence

$$-1 < \mu_{ic} < 0 < \mu_{oc} < \mu_r < \mu_e$$

by rules that we formally describe in Algorithm `nelder`. Our formulation of the algorithm allows for termination if either $f(x_{N+1}) - f(x_1)$ is sufficiently small or a user-specified number of function evaluations has been expended.

ALGORITHM 8.1.1. `nelder`$(S, f, \tau, kmax)$

1. *Evaluate f at the vertices of S and sort the vertices of S so that* (8.1) *holds.*

2. *Set $fcount = N + 1$.*

3. *While $f(x_{N+1}) - f(x_1) > \tau$*

   (a) *Compute $\overline{x}$,* (8.3), *$x(\mu_r)$,* (8.2), *and $f_r = f(x(\mu_r))$. $fcount = fcount + 1$.*

   (b) **Reflect:** *If $fcount = kmax$ then exit. If $f(x_1) \le f_r < f(x_N)$, replace x_{N+1} with $x(\mu_r)$ and go to step* 3g.

   (c) **Expand:** *If $fcount = kmax$ then exit. If $f_r < f(x_1)$ then compute $f_e = f(x(\mu_e))$. $fcount = fcount + 1$. If $f_e < f_r$, replace x_{N+1} with $x(\mu_e)$; otherwise replace x_{N+1} with $x(\mu_r)$. Go to to step* 3g.

   (d) **Outside Contraction:** *If $fcount = kmax$ then exit. If $f(x_N) \le f_r < f(x_{N+1})$, compute $f_c = f(x(\mu_{oc}))$. $fcount = fcount + 1$. If $f_c \le f_r$ replace x_{N+1} with $x(\mu_{oc})$ and go to step* 3g*; otherwise go to step* 3f.

   (e) **Inside Contraction:** *If $fcount = kmax$ then exit. If $f_r \ge f(x_{N+1})$ compute $f_c = f(x(\mu_{ic}))$. $fcount = fcount + 1$. If $f_c < f(x_{N+1})$, replace x_{N+1} with $x(\mu_{ic})$ and go to step* 3g*; otherwise go to step* 3f.

   (f) **Shrink:** *If $fcount \ge kmax - N$, exit. For $2 \le i \le N + 1$: set $x_i = x_1 - (x_i - x_1)/2$; compute $f(x_i)$.*

   (g) **Sort:** *Sort the vertices of S so that* (8.1) *holds.*

A typical sequence [169] of candidate values for μ is

$$\{\mu_r, \mu_e, \mu_{oc}, \mu_{ic}\} = \{1, 2, 1/2, -1/2\}.$$

Figure 8.1 is an illustration of the options in two dimensions. The vertices labeled 1, 2, and 3 are those of the original simplex.

The Nelder–Mead algorithm is not guaranteed to converge, even for smooth problems [89], [188]. The failure mode is stagnation at a nonoptimal point. In §8.1.3 we will present some examples from [188] that illustrate this failure. However, the performance of the Nelder–Mead algorithm in practice is generally good [169], [274]. The shrink step is rare in practice and we will assume in the analysis in §8.1.2 that shrinks do not occur. In that case, while a Nelder–Mead iterate may not result in a reduction in the best function value, the average value

$$\underline{f} = \frac{1}{N+1} \sum_{j=1}^{N+1} f(x_j)$$

will be reduced.

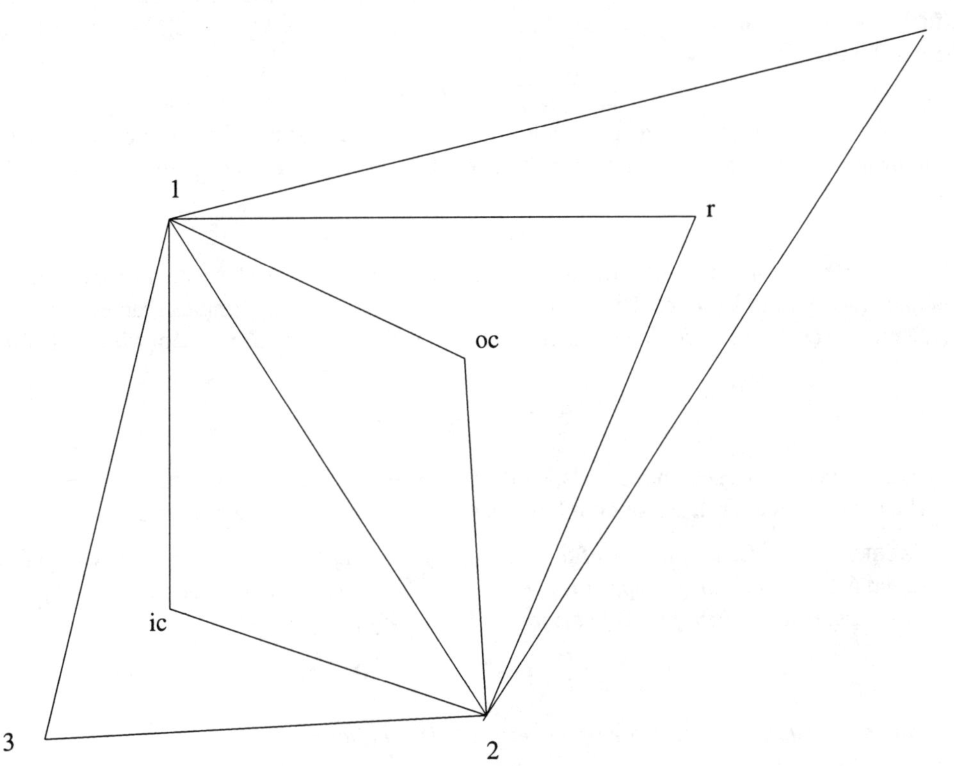

Figure 8.1: *Nelder–Mead Simplex and New Points*

8.1.2 Sufficient Decrease and the Simplex Gradient

Our study of the Nelder–Mead algorithm is based on the simple ideas in §3.1. We will denote the vertices of the simplex S^k at the kth iteration by $\{x_j^k\}_{j=1}^{N+1}$. We will simplify notation by suppressing explicit mention of S^k in what follows by denoting

$$V^k = V(S^k), \delta^k = \delta(f : S^k), K^k = K(S^k), \text{ and } D^k(f) = D(f : S^k).$$

If V^0 is nonsingular then V^k is nonsingular for all $k > 0$ [169]. Hence if S^0 is nonsingular so is S^k for all k and hence $D^k(f)$ is defined for all k.

We formalize this by assuming that our sequence of simplices satisfies the following assumption.

ASSUMPTION 8.1.1. *For all k,*

- S^k *is nonsingular.*
- *The vertices satisfy* (8.1).
- $\underline{f}^{k+1} < \underline{f}^k$.

Assumption 8.1.1 is satisfied by the Nelder–Mead sequence if no shrink steps are taken and the initial simplex directions are linearly independent [169]. The Nelder–Mead algorithm demands that the average function value improve, but no control is possible on which value is improved, and the simplex condition number can become unbounded.

We can define a sufficient decrease condition for search algorithms that is analogous to the sufficient decrease condition for steepest descent and related algorithms (3.2). We will ask that the $k+1$st iteration satisfy

$$\underline{f}^{k+1} - \underline{f}^k < -\alpha \|D^k f\|^2. \tag{8.4}$$

Here $\alpha > 0$ is a small parameter. Our choice of sufficient decrease condition is motivated by the smooth case and steepest descent, where (3.2) and the lower bound $-\bar{\lambda}$ on λ from Lemma 3.2.3 lead to

$$f(x_{k+1}) - f(x_k) \le -\bar{\lambda}\alpha \|\nabla f(x_k)\|^2,$$

which is a smooth form of (8.4). Unlike the smooth case, however, we have no descent direction and must incorporate $\bar{\lambda}$ into α. This leads to the possibility that if the simplex diameter is much smaller than $\|D^k f\|$, (8.4) could fail on the first iterate. We address this problem with the scaling

$$\alpha = \alpha_0 \frac{\sigma_+(S^0)}{\|D^0 f\|}.$$

A typical choice in line search methods, which we use in our numerical results, is $\alpha_0 = 10^{-4}$.

The convergence result for smooth functions follows easily from Lemma 6.2.1.

THEOREM 8.1.1. *Let a sequence of simplices satisfy Assumption* 8.1.1 *and let the assumptions of Lemma* 6.2.1 *hold, with the Lipschitz constants K^k uniformly bounded. Assume that $\{\underline{f}^k\}$ is bounded from below. Then if* (8.4) *holds for all but finitely many k and*

$$\lim_{k\to\infty} \sigma_+(S^k)\kappa(V^k) = 0,$$

then any accumulation point of the simplices is a critical point of f.

Proof. The boundedness from below of $\{\underline{f}^k\}$ and (8.4) imply that $f^k \to 0$. Assumption 8.1.1 and (8.4) imply that $\lim_{k\to\infty} D^k f = 0$. Hence (6.2) implies

$$\lim_{k\to\infty} \|\nabla f(x_1^k)\| \le \lim_{k\to\infty} \left(K\kappa(V^k)\sigma_+(S^k) + \|D^k f\|\right) = 0.$$

Hence, if x^* is any accumulation point of the sequence $\{x_1^k\}$ then $\nabla f(x^*) = 0$. This completes the proof since $\kappa(V^k) \ge 1$ and therefore $\sigma_+(V^k) \to 0$. □

The result for the noisy functions that satisfy (6.1) with f_s smooth reflects the fact that the resolution is limited by the size of ϕ. In fact, if $\sigma_+(S^k)$ is much smaller than $\|\phi\|_{S^k}$, no information on f_s can be obtained by evaluating f at the vertices of S^k and once $\sigma_+(S^k)$ is smaller than $\|\phi\|_{S^k}^{1/2}$ no conclusions on ∇f_s can be drawn. If, however, the noise decays to zero sufficiently rapidly near the optimal point, the conclusions of Theorem 8.1.1 still hold.

THEOREM 8.1.2. *Let a sequence of simplices satisfy Assumption* 8.1.1 *and let the assumptions of Lemma* 6.2.2 *hold with the Lipschitz constants K_s^k uniformly bounded. Assume that $\{\underline{f}^k\}$ is bounded from below. Then if* (8.4) *holds for all but finitely many k and if*

$$\lim_{k\to\infty} \kappa(V^k)\left(\sigma_+(S^k) + \frac{\|\phi\|_{S^k}}{\sigma_+(S^k)}\right) = 0, \tag{8.5}$$

then any accumulation point of the simplices is a critical point of f_s.

Proof. Our assumptions, as in the proof of Theorem 8.1.1, imply that $D^k f \to 0$. Recall that Lemma 6.2.2 implies that

$$\|D^k f_s\| \le \|D^k f\| + K^k \kappa(V^k)\left(\sigma_+(S^k) + \frac{\|\phi\|_{S^k}}{\sigma_+(S^k)}\right), \tag{8.6}$$

and the sequence $\{K^k\}$ is bounded because $\{K_s^k\}$ is. Hence, by (8.5), $D^k f_s \to 0$ as $k \to \infty$. □

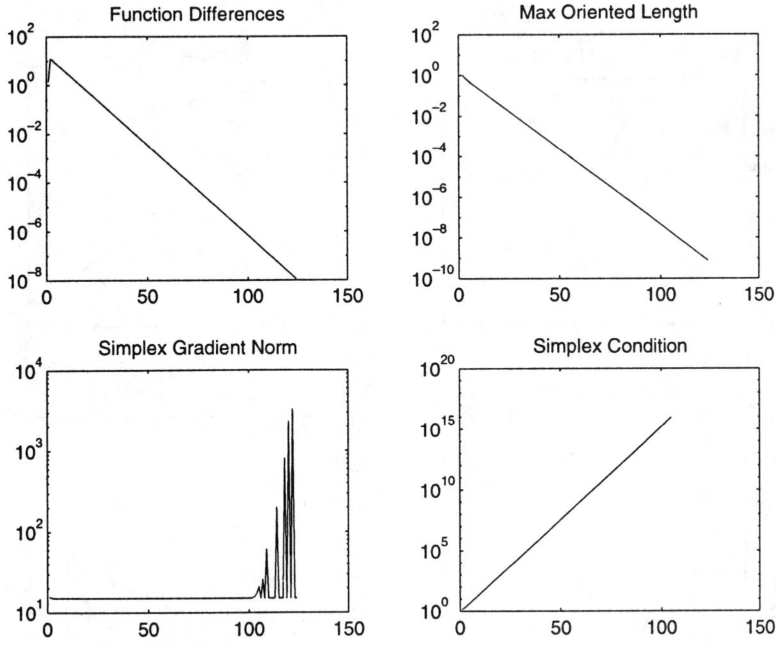

Figure 8.2: *Unmodified Nelder–Mead,* $(\tau, \theta, \phi) = (1, 15, 10)$

8.1.3 McKinnon's Examples

In this set of three examples from [188], $N = 2$, and

$$f(x) = \begin{cases} \theta\phi|(x)_1|^\tau + (x)_2 + (x)_2^2, & (x)_1 \le 0, \\ \\ \theta(x)_1^\tau + (x)_2 + (x)_2^2, & (x)_1 > 0. \end{cases}$$

The examples in [188] consider the parameter sets

$$(\tau, \theta, \phi) = \begin{cases} (3, 6, 400), \\ (2, 6, 60), \\ (1, 15, 10). \end{cases}$$

The initial simplex was

$$x_1 = (1,1)^T, x_2 = (\lambda_+, \lambda_-)^T, x_3 = (0,0)^T, \text{ where } \lambda_\pm = (1 \pm \sqrt{33})/8.$$

With this data, the Nelder–Mead iteration will stagnate at the origin, which is not a critical point for f. The stagnation mode is repeated inside contractions that leave the best point (which is not a minimizer) unchanged.

We terminated the iteration when the difference between the best and worst function values was $< 10^{-8}$.

We illustrate the behavior of the Nelder–Mead algorithm in Figures 8.2, 8.3, and 8.4. In all the figures we plot, as functions of the iteration index, the difference between the best and worst function values, σ_+, the maximum oriented length, the norm of the simplex gradient, and the l^2 condition number of the matrix of simplex directions. In all three problems stagnation is evident from the behavior of the simplex gradients. Note also how the simplex condition number is growing rapidly.

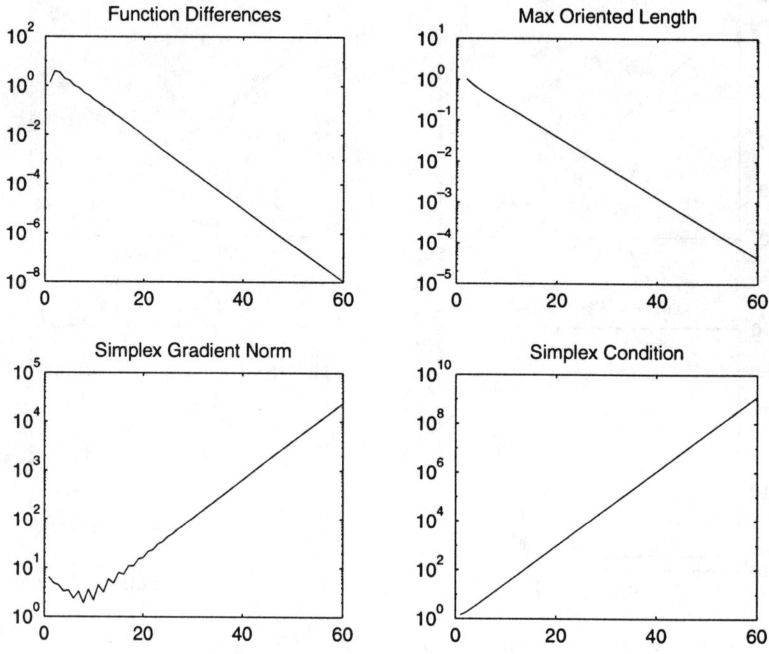

Figure 8.3: *Unmodified Nelder–Mead,* $(\tau, \theta, \phi) = (2, 6, 60)$

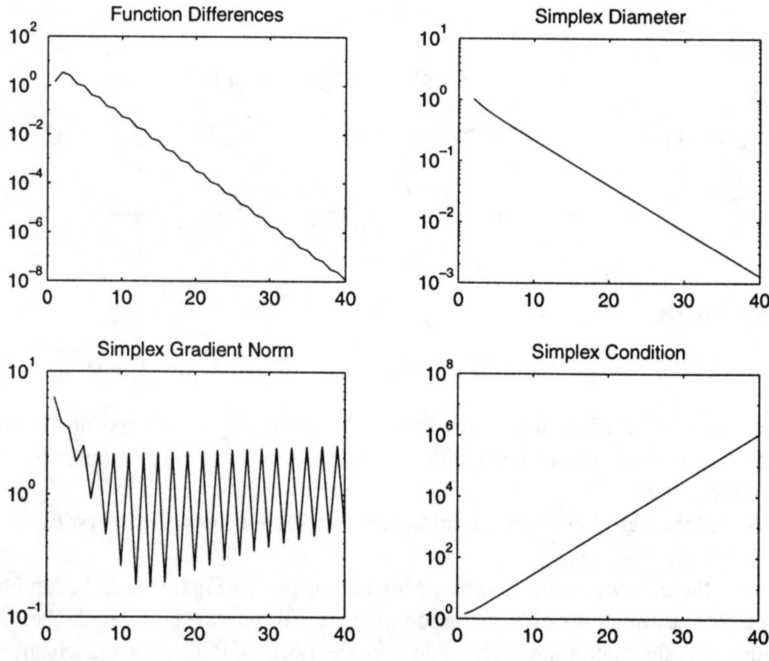

Figure 8.4: *Unmodified Nelder–Mead,* $(\tau, \theta, \phi) = (3, 6, 400)$

8.1.4 Restarting the Nelder–Mead Algorithm

When the Nelder–Mead iteration stagnates, a restart with the same best point and a different set of directions can help sometimes. In order to formulate a restart scheme, one must first develop a strategy for detecting stagnation. One might think that a large simplex condition would suffice for this. However [204], the ability of the Nelder–Mead simplices to drastically vary their shape is an important feature of the algorithm and looking at the simplex condition alone would lead to poor results. Failure of (8.4), however, seems to indicate that something is wrong, and we will use that as our stagnation detector.

Having detected stagnation, one must modify the simplex. Simply performing a shrink step is not effective. The method we advocate here, from [155], is the *oriented restart.* The motivation is that if the simplex gradient can be trusted to be in the correct orthant in R^N, a new, smaller simplex with orthogonal edges oriented with that quadrant should direct the iteration in a productive direction.

We propose performing an *oriented restart* when (8.4) fails but $\underline{f}^{k+1} - \underline{f}^k < 0$. This means replacing the current simplex with vertices $\{x_j\}_{j=1}^{N+1}$, ordered so that (8.1) holds, with a new smaller simplex having vertices (before ordering!) $\{y_j\}_{j=1}^{N+1}$ with $y_1 = x_1$ and

$$y_j = y_1 - \beta_{j-1} e_{j-1} \text{ for } 2 \leq j \leq N+1, \tag{8.7}$$

where, for $1 \leq l \leq N$, e_l is the lth coordinate vector,

$$\beta_l = \frac{1}{2} \begin{cases} \sigma_-(S^k)\mathrm{sign}((D^k f)_l), & (D^k f)_l \neq 0, \\ \sigma_-(S^k), & (D^k f)_l = 0, \end{cases}$$

and $(D^k f)_l$ is the lth component of $D^k f$. If $D^k f = 0$ we assume that the Nelder–Mead iteration would have been terminated at iteration k because there is no difference between best and worst values.

So, before ordering, the new simplex has the same first point as the old. The diameter of the new simplex has not been increased since the diameter of the new simplex is at most $\sigma_+(S^k)$. Moreover all edge lengths have been reduced. So after reordering $\sigma_+(S^{k+1}) \leq \sigma_-(S^k)$. As for κ, after the oriented shrink, but before reordering, $\kappa(V) = 1$. After reordering, of course, the best point may no longer be x_1. In any case the worst-case bound on κ is

$$\kappa(V^{k+1}) = \|V^{k+1}\|^2 \leq (1 + \sqrt{N})^2. \tag{8.8}$$

In any case, the new simplex is well conditioned.

Returning to the McKinnon examples, we find that an oriented restart did remedy stagnation for the smooth examples. The graphs in Figures 8.5, 8.6, and 8.7 report the same data as for the unmodified algorithm, with stars on the plots denoting oriented restarts.

For the smoothest example, $(\tau, \theta, \phi) = (3, 6, 400)$, the modified form of Nelder–Mead took a single oriented restart at the 21st iteration. For the less smooth of these two, $(\tau, \theta, \phi) = (2, 6, 60)$, a single restart was taken on the 19th iteration. As one can see from Figures 8.6 and 8.7 the restart had an immediate effect on the simplex gradient norm and overcame the stagnation.

For the nonsmooth example, $(\tau, \theta, \phi) = (1, 15, 10)$, in Figure 8.5, the modified algorithm terminated with failure after restarting on the 44th, 45th, and 46th iterations. Since the objective is not smooth at the stagnation point, this is the best we can expect and is far better than the behavior of the unmodified algorithm, which stagnates with no warning of the failure.

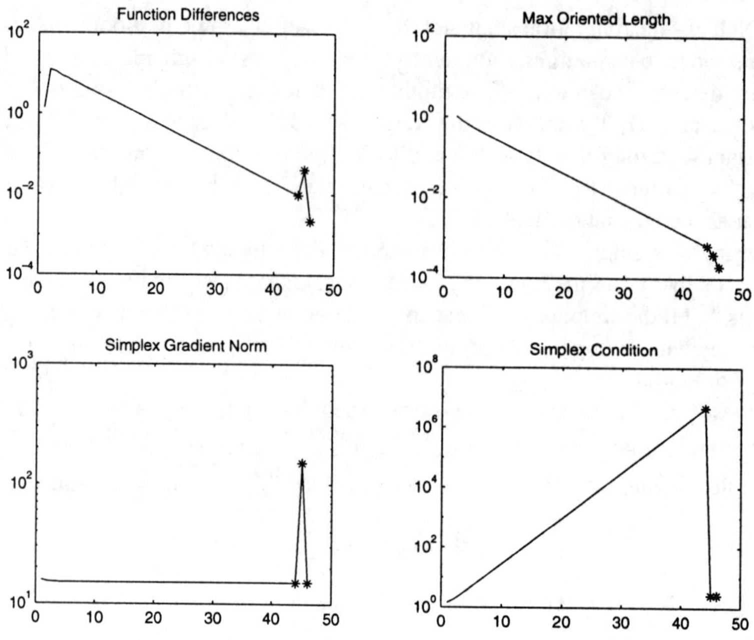

Figure 8.5: *Modified Nelder–Mead,* $(\tau, \theta, \phi) = (1, 15, 10)$

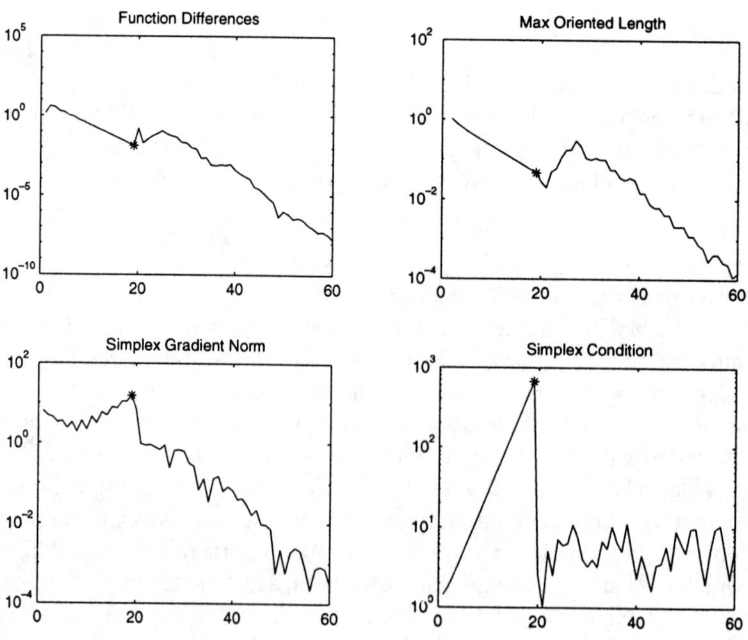

Figure 8.6: *Modified Nelder–Mead,* $(\tau, \theta, \phi) = (2, 6, 60)$

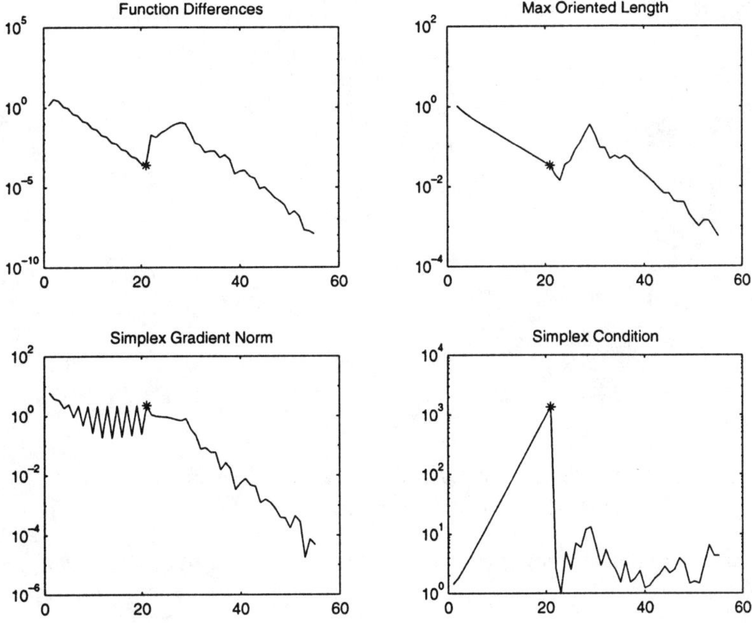

Figure 8.7: *Modified Nelder–Mead,* $(\tau, \theta, \phi) = (3, 6, 400)$

8.2 Multidirectional Search

8.2.1 Description and Implementation

One way to address the possible ill-conditioning in the Nelder–Mead algorithm is to require that the condition numbers of the simplices be bounded. The multidirectional search (MDS) method [85], [260], [261] does this by making each new simplex congruent to the previous one. The results in this section, mostly taken from [29], show that MDS has convergence properties like those of implicit filtering.

In the special case of equilateral simplices, V^k is a constant multiple of V^0 and the simplex condition number is constant. If the simplices are not equilateral, then $\kappa(V)$ may vary depending on which vertex is called x_1, but (6.6) will hold in any case.

Figure 8.8 illustrates the two-dimensional case for two types of simplices. Beginning with the ordered simplex S^c with vertices x_1, x_2, x_3 one first attempts a *reflection* step, leading to a simplex S^r with vertices x_1, r_2, r_3.

If the best function value of the vertices of S^r is better than the best $f(x_1)$ in S^0, S^r is (provisionally) accepted and *expansion* is attempted. The expansion step differs from that in the Nelder–Mead algorithm because N new points are needed to make the new, larger simplex similar to the old one. The expansion simplex S^e has vertices x_1, e_2, e_3 and is accepted over S^r if the best function value of the vertices of S^e is better than the best in S^r. If the best function value of the vertices of S^r is not better than the best in S^c, then the simplex is *contracted* and the new simplex has vertices x_1, c_2, c_3. After the new simplex is identified, the vertices are reordered to create the new ordered simplex S^+.

Similar to the Nelder–Mead algorithm, there are expansion and contraction parameters μ_e and μ_c. Typical values for these are 2 and $1/2$.

ALGORITHM 8.2.1. $\texttt{mds}(S, f, \tau, kmax)$

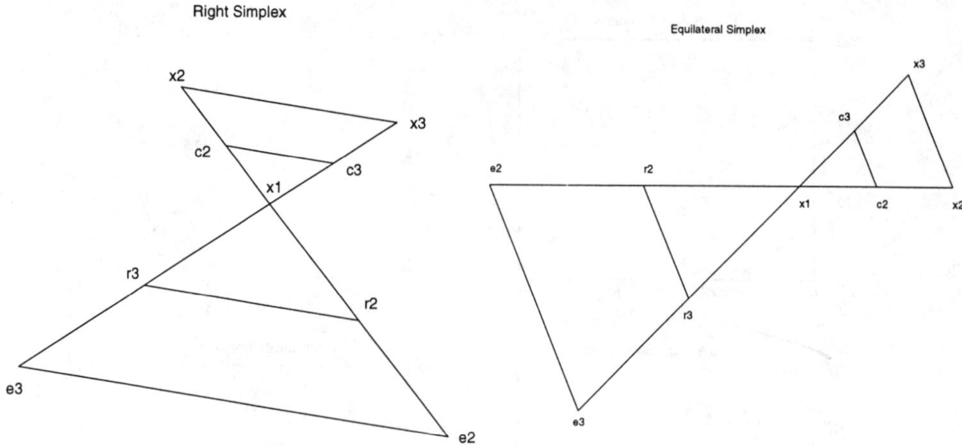

Figure 8.8: *MDS Simplices and New Points*

1. *Evaluate f at the vertices of S and sort the vertices of S so that* (8.1) *holds.*
2. *Set $fcount = N + 1$.*
3. *While $f(x_{N+1}) - f(x_1) > \tau$*
   (a) **Reflect:** *If $fcount = kmax$ then exit.*
   For $j = 2, \ldots, N+1$: $r_j = x_1 - (x_j - x_1)$; Compute $f(r_j)$; $fcount = fcount + 1$.
   If $f(x_1) > \min_j\{f(r_j)\}$ then goto step 3b *else goto step* 3c.
   (b) **Expand:**
      i. *For $j = 2, \ldots, N + 1$: $e_j = x_1 - \mu_e(x_j - x_1)$; Compute $f(e_j)$; $fcount = fcount + 1$.*
      ii. *If $\min_j\{f(r_j)\} > \min_j\{f(e_j)\}$ then*
      for $j = 2, \ldots N + 1$: $x_j = e_j$
      else
      for $j = 2, \ldots N + 1$: $x_j = r_j$
      iii. *Goto step* 3d
   (c) **Contract:** *For $j = 2, \ldots, N + 1$: $x_j = x_1 + \mu_c(x_j - x_1)$, Compute $f(x_j)$*
   (d) **Sort:** *Sort the vertices of S so that* (8.1) *holds.*

If the function values at the vertices of S^c are known, then the cost of computing S^+ is $2N$ additional evaluations. Just as with the Nelder–Mead algorithm, the expansion step is optional but has been observed to improve performance.

The extension of MDS to bound constrained and linearly constrained problems is not trivial. We refer the reader to [173] and [175] for details.

8.2.2 Convergence and the Simplex Gradient

Assume that the simplices are either equilateral or right simplices (having one vertex from which all N edges are at right angles). In those cases, as pointed out in [262], the possible vertices created by expansion and reflection steps form a regular lattice of points. If the MDS simplices remain bounded, only finitely many reflections and expansions are possible before every point on that lattice has been visited and a contraction to a new maximal simplex size must take place.

This exhaustion of a lattice takes place under more general conditions [262] but is most clear for the equilateral and right simplex cases.

Theorem 6.2.9 implies that infinitely many contractions and convergence of the simplex diameters to zero imply convergence of the simplex gradient to zero. The similarity of Theorem 6.2.9 to Lemma 6.2.2 and of Theorem 8.2.1, the convergence result for multidirectional search, to Theorem 8.1.2 is no accident. The Nelder–Mead iteration, which is more aggressive than the multidirectional search iteration, requires far stronger assumptions (well conditioning and sufficient decrease) for convergence, but the ideas are the same. Theorems 6.2.9 and 8.2.1 can be used to extend the results in [262] to the noisy case. The observation in [85] that one can apply any heuristic or machine-dependent idea to improve performance, say, by exploring far away points on spare processors (the "speculative function evaluations" of [46]) without affecting the analysis is still valid here.

THEOREM 8.2.1. *Let f satisfy* (6.1) *and assume that the set*

$$\{x \mid f(x) \le f(x_1^0)\}$$

is bounded. Assume that the simplex shape is such that

$$\lim_{k\to\infty} \sigma_+(S^k) \to 0. \tag{8.9}$$

Let B^k be a ball of radius $2\sigma_+(S^k)$ about x_1^k. Then if

$$\lim_{k\to\infty} \frac{\|\phi\|_{B^k}}{\sigma_+(S^k)} = 0$$

then every limit point of the vertices is a critical point of f_s.

Recall that if the simplices are equilateral or right simplices, then (8.9) holds (see exercise 8.6.2).

8.3 The Hooke–Jeeves Algorithm

8.3.1 Description and Implementation

The *Hooke–Jeeves algorithm* is like implicit filtering in that the objective is evaluated on a stencil and the function values are used to compute a search direction. However, unlike implicit filtering, there are only finitely many possible search directions and only qualitative information about the function values is used.

The algorithm begins with a *base point* x and pattern size h, which is like the scale in implicit filtering. In the next phase of the algorithm, called the *exploratory move* in [145], the function is sampled at successive perturbations of the base point in the search directions $\{v_j\}$, where v_j is the jth column of a direction matrix V. In [145] and our MATLAB implementation $V = I$. The current best value $f_{cb} = f(x_{cb})$ and best point x_{cb} are recorded and returned. x_{cb} is initialized to x. The sampling is managed by first evaluating f at $x_{cb} + v_j$ and only testing $x_{cb} - v_j$ if $f(x_{cb} + v_j) \ge f(x_{cb})$. The exploratory phase will either produce a new base point or fail (meaning that $x_{cb} = x$). Note that this phase depends on the ordering of the coordinates of x. Applying a permutation to x could change the output of the exploration.

If the exploratory phase has succeeded, the search direction is

$$d^{HJ} = x_{cb} - x \tag{8.10}$$

and the new base point is x_{cb}. The subtle part of the algorithm begins here. Rather than center the next exploration at x_{cb}, which would use some of the same points that were examined in the previous exploration, the Hooke–Jeeves *pattern move* step is aggressive and tries to move further. The algorithm centers the next exploratory move at

$$x_C = x + 2d^{HJ} = x_{cb} + d^{HJ}.$$

If this second exploratory move fails to improve upon $f(x_{cb})$, then an exploratory move with x_{cb} as the center is tried. If that fails h is reduced, x is set to x_{cb}, and the process is started over. Note that when h has just been set, the base point and the center of the stencil for the exploratory moves are the same, but afterward they are not.

If, after the first exploratory move, $x_{cb} = x$ (i.e., as it will be if x is the best point in the pattern), then x is left unchanged and h is reduced.

Therefore, whenever h is reduced, the stencil centered at x has x itself as the best point. This is exactly the situation that led to a shrink in the MDS algorithm and, as you might expect, will enable us to prove a convergence result like those in the previous sections. In [145] h was simply multiplied by a constant factor. Our description in Algorithm `hooke` follows the model of implicit filtering and uses a sequence of scales. Choice of perturbation directions could be generalized to any simplex shape, not just the right simplices used in [145].

Figure 8.9 illustrates the idea for $N = 2$. The base point x lies at the center of the stencil. If

$$f(x_1^+) < f(x), f(x_2^+) < f(x), f(x_1^-) \geq f(x), \text{ and } f(x_2^-) \geq f(x),$$

then the new base point x_b will be located above and to the right of x. The next exploratory move will be centered at x_C, which is the center of the stencil in the upper right corner of the figure.

The reader, especially one who plans to implement this method, must be mindful that points may be sampled more than once. For example, in the figure, if the exploratory move centered at x_C fails, f will be evaluated for the second time at the four points in the stencil centered at x_b unless the algorithm is implemented to avoid this. The MDS method is also at risk of sampling points more than once. The implementations of Hooke–Jeeves and MDS in our suite of MATLAB codes keep the most recent $4N$ iterations in memory to guard against this. This reevaluation is much less likely for the Nelder–Mead and implicit filtering methods. One should also be aware that the Hooke–Jeeves algorithm, like Nelder–Mead, does not have the natural parallelism that implicit filtering and MDS do.

One could implement a variant of the Hooke–Jeeves iteration by using $x_C = x + d^{HJ}$ instead of $x_C = x + 2d^{HJ}$ and shrinking the size of the simplex on stencil failure. This is the discrete form of the classical *coordinate descent algorithm* [180] and can also be analyzed by the methods of this section (see [279] for a different view).

Our implementation follows the model of implicit filtering as well as the description in [145]. We begin with the exploratory phase, which uses a base point x_b, base function value $f_b = f(x_b)$, and stencil center x_C. Note that in the algorithm $x_b = x_C$ for the first exploration and $x_C = x_b + d^{HJ}$ thereafter. Algorithm `hjexplore` takes a base point and a scale and returns a direction and the value at the trial point $x + d$. We let $V = I$ be the matrix of coordinate directions, but any nonsingular matrix of search directions could be used. The status flag s_f is used to signal failure and trigger a shrink step.

ALGORITHM 8.3.1. `hjexplore`(x_b, x_C, f, h, s_f)

1. $f_b = f(x_b)$; $d = 0$; $s_f = 0$; $x_{cb} = x_b$; $f_{cb} = f(x_b)$; $x_t = x_C$

2. *for* $j = 1, \dots, N$: $p = x_t + hv_j$; *if* $f(p) \geq f_b$ *then* $p = x_t - hv_j$;
   if $f(p) < f_b$ *then* $x_t = x_{cb} = p$; $f_b = f(x_{cb})$

3. *if* $x_{cb} \neq x_b$; $s_f = 1$; $x_b = x_{cb}$

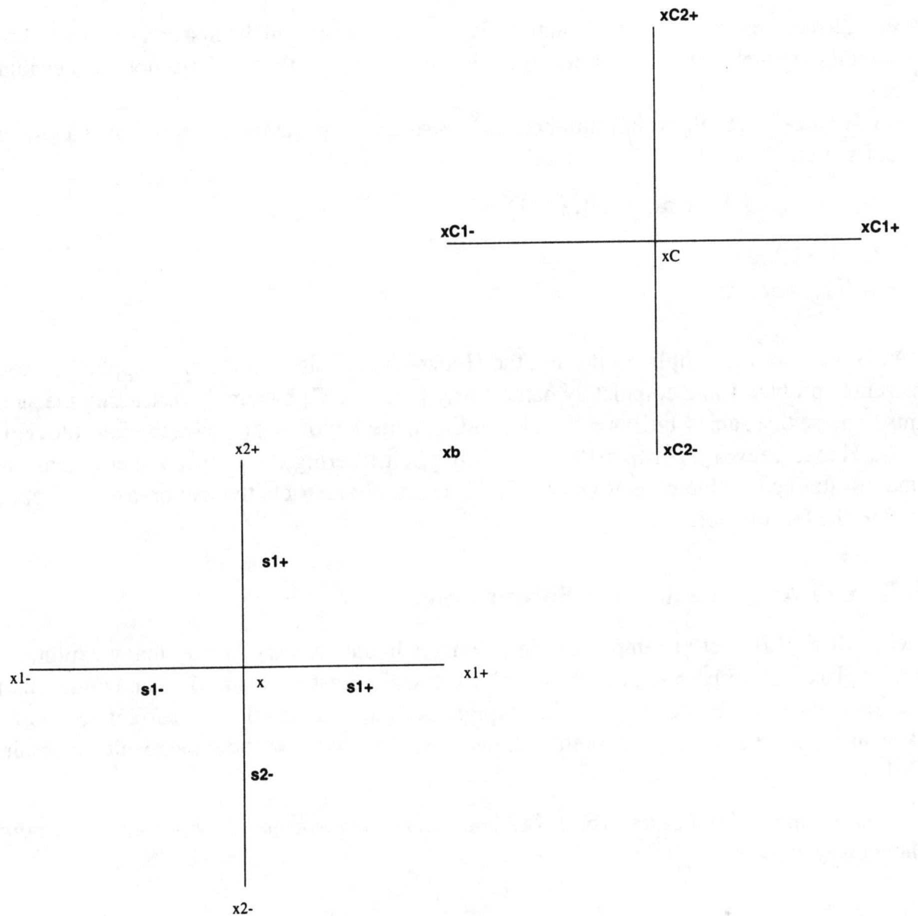

Figure 8.9: *Hooke–Jeeves Pattern and New Points*

The exploration is coupled to the pattern move to complete the algorithm for a single value of the scale. The inputs for Algorithm `hjsearch` are an initial iterate x, the function, and the scale. On output, a point x is returned for which the exploration has failed. There are other considerations, such as the budget for function evaluations, that should trigger a return from the exploratory phase in a good implementation. In our MATLAB code `hooke.m` we pay attention to the number of function evaluations and change in the function value as part of the decision to return from the exploratory phase.

ALGORITHM 8.3.2. `hjsearch`(x, f, h)

1. $x_b = x$; $x_C = x$; $s_f = 1$

2. *Call* `hjexplore`(x, x_C, f, h, s_f)

3. *While* $s_f = 1$

   (a) $d = x - x_b$; $x_b = x$; $x_C = x + d$

   (b) *Call* `hjexplore`(x, x_C, f, h, s_f);
   If $s_f = 0$; $x_C = x$; *Call* `hjexplore`(x, x_C, f, h, s_f)

Step 3b requires care in implementation. If $s_f = 0$ on exit from the first call to `hjexplore`, one should only test f at those points on the stencil centered at x that have not been evaluated before.

The Hooke–Jeeves algorithm simply calls `hjsearch` repeatedly as h varies over a sequence $\{h_k\}$ of scales.

ALGORITHM 8.3.3. `hooke`$(x, f, \{h_k\})$

1. *For* $k = 1, \ldots$
   Call `hjsearch`(x, f, h_k)

As is the case with implicit filtering, the Hooke–Jeeves algorithm can be applied to bound constrained problems in a completely natural way [145], [227] by simply restricting the stencil points to those that satisfy the bounds and avoiding pattern moves that leave the feasible region.

The Hooke–Jeeves algorithm shares with implicit filtering the property that extension to bound constrained problems is trivial [145]. One simply restricts the exploratory and pattern moves to the feasible set.

8.3.2 Convergence and the Simplex Gradient

As with MDS, if the set of sampling points remains bounded, only finitely many explorations can take place before `hjsearch` returns and the scale must be reduced. The conditions for reduction in the scale include failure of an exploratory move centered at the current best point x. This means that we can apply Theorem 6.2.9 with $\kappa_+ = 1$ to prove the same result we obtained for MDS.

THEOREM 8.3.1. *Let f satisfy* (6.1). *Let $\{x_k\}$ be the sequence of Hooke–Jeeves best points. Assume that the set*

$$\{x \,|\, f(x) \leq f(x_0)\}$$

is bounded. Then let $h_k \to 0$ and if

$$\lim_{k\to\infty} \frac{\|\phi\|_{B^k}}{\sigma_+(S^k)} = 0,$$

where B^k is the ball of radius $2h_k$ about x_k, then every limit point of $\{x_k\}$ is a critical point of f_s.

8.4 Other Approaches

In this section we briefly discuss two methods that have been used successfully for noisy problems. These methods are substantially more difficult to implement than the ones that we have discussed so far and we will give few details. The pointers to the literature are a good starting place for the interested and energetic reader.

8.4.1 Surrogate Models

As any sampling method progresses, the function values can be used to build a (possibly) quadratic model based, for example, on interpolation or least squares fit-to-data. Such models are called *surrogates* or *response surfaces*. Even for smooth f there are risks in doing this. Points from early in the iteration may corrupt an accurate model that could be built from the more recent points; however, the most recent points alone may not provide a rich enough set of interpolatory data. The function being modeled could be too complex to be modeled in a simple way (think of the Lennard–Jones function), and very misleading results could be obtained. However, this approach is often very productive even for smooth problems in which evaluation of f is very expensive (see [28] for a high-flying example).

Initialization of the model requires an initial set of points at which to sample f. Selection of this point set is not a trivial issue, and the regular stencils used in implicit filtering and the direct search algorithms are very poor choices. The study of this issue alone is a field in itself, called *design and analysis of computer experiments* (DACE) [27], [167], [230].

Having built such a model, one then finds one or more local minima of the model. One can use either a conventional gradient-based method, a sampling algorithm of the type discussed in Chapters 7 or 8, or an algorithm that is itself based on building models like the one described in [62], the nongradient-based approaches being used when the model is expected to be multimodal or nonconvex. Upon minimizing the model, one then evaluates f again at one or more new points.

The implementation of such a scheme requires careful coordination between the sampling of the function, the optimization of the model, and the changing of the set of sample points. We refer the reader to [28] and [4] for more information on recent progress in this area.

8.4.2 The DIRECT Algorithm

Suppose f is a Lipschitz continuous function on $[a, b]$ with Lipschitz constant L. If one has a priori knowledge of L, one can use this in a direct search algorithm to eliminate intervals of possible optimal points based on the function values at the endpoints of these intervals. The *Shubert algorithm* [146], [214], [241] is the simplest way to use this idea. The method begins with the fact that

$$f(x) \geq f_{low}(x, a, b) = \max(f(a) - L(x - a), f(b) - L(b - x)) \tag{8.11}$$

for all $x \in [a, b]$. If one samples f repeatedly, one can use (8.11) on a succession of intervals and obtain a piecewise linear approximation to f. If $I_n = [a_n, b_n] \subset [a, b]$ then $f(x) \geq f_{low}(x, a_n, b_n)$ on I_n, the minimum value of $f_{low}(x, a_n, b_n)$ is

$$V_n = (f(a_n) + f(b_n) - L(b_n - a_n))/2,$$

and the minimizer is

$$M_n = (f(a_n) - f(b_n) + L(b_n + a_n))/(2L).$$

The algorithm begins with $I_0 = [a, b]$, selects the interval for which V_n is least, and divides at M_n. This means that if K intervals have been stored we have, replacing I_n and adding I_{K+1} to the list,

$$I_n = [a_n, M_n] \text{ and } I_{K+1} = [M_n, b_n].$$

The sequence of intervals is only ordered by the iteration counter, not by location. In this way the data structure for the intervals is easy to manage.

If there are p and k such that $p \neq k$ and $V_p \geq \max(f(a_k), f(b_k))$, then I_p need not be searched any longer, since the best value from I_p is worse than the best value in I_k. The algorithm's rule for division automatically incorporates this information and will never sample from I_p.

There are two problems with this algorithm. One cannot expect to know the Lipschitz constant L, so it must be estimated. An estimated Lipschitz constant that is too low can lead to erroneous rejection of an interval. An estimate that is too large will lead to slow convergence, since intervals that should have been discarded will be repeatedly divided. The second problem

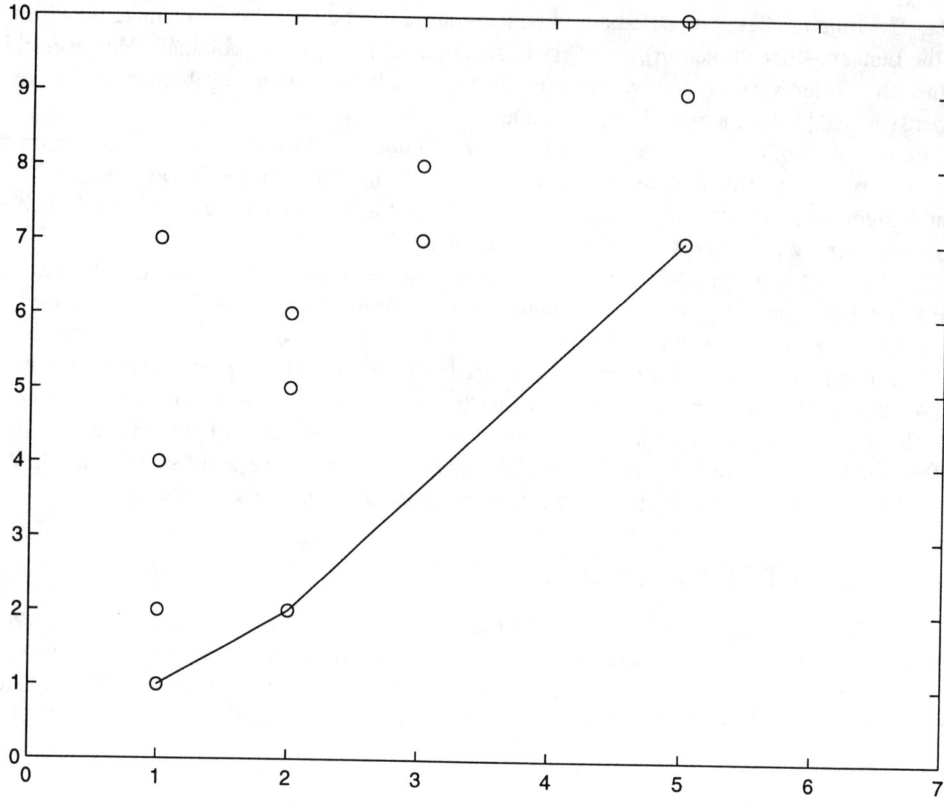

Figure 8.10: *Selection of Intervals in DIRECT*

is far more serious. The obvious generalization of the Shubert algorithm to more than one dimension would replace intervals by N-dimensional hyperrectangles and require sampling at each of the 2^N vertices of the rectangle to be divided. This exponential complexity makes this trivial generalization of the Shubert algorithm completely impractical.

The DIRECT algorithm [150] attempts to address these problems by sampling at the midpoint of the hyperrectangle rather than the vertices and indirectly estimating the Lipschitz constant as the optimization progresses. The scheme is not completely successful in that the mesh of sample points becomes everywhere dense as the optimization progresses. Hence the algorithm becomes an exhaustive search, a fact that is used in [150] to assert global convergence. In spite of the exponential complexity of exhaustive search, even one with a fixed-size mesh (a problem with any deterministic algorithm that is truly global [248]), DIRECT has been reported to perform well in the early phases of the iteration [150], [108] and for suites of small test problems. DIRECT is worth consideration as an intermediate algorithmic level between methods like implicit filtering, Nelder–Mead, Hooke–Jeeves, or MDS on the conservative side and nondeterministic methods like simulated annealing or genetic algorithms on the radical side.

We will describe DIRECT completely only for the case $N = 1$. This will make clear how the algorithm implicitly estimates the Lipschitz constant. The extension to larger values of N requires careful management of the history of subdivision of the hyperrectangles, and we will give a simple pictorial account of that. For more details we refer to [150], [147], or the documentation [108] of the FORTRAN implementation of DIRECT from the software collection.

As with the Shubert algorithm we begin with an interval $[a, b]$ but base our lower bound and our subdivision strategy on the midpoint $c = (a + b)/2$. If the Lipschitz constant L is known then

$$f(x) \geq f(c) - L(b-a)/2.$$

If we are to divide an interval and also retain the current value c as the midpoint of an interval in the set of intervals, we must divide an interval into three parts. If there are K intervals on the list and an interval $I_n = [a_n, b_n]$ with midpoint c_n has been selected for division, the new intervals are

$$I_{K+1} = [a_n, a_n + (b_n - a_n)/3], I_n = [a_n + (b_n - a_n)/3, b_n - (b_n - a_n)/3], \text{ and}$$
$$I_{K+2} = [b_n - (b_n - a_n)/3, b_n].$$

So c_n is still the midpoint of I_n and two new midpoints have been added.

The remaining part of the algorithm is the estimation of the Lipschitz constant and the simultaneous selection of the intervals to be divided. If the Lipschitz constant were known, an interval would be selected for division if $f(c) - L(b-a)/2$ were smallest. This is similar to the Shubert algorithm. In order for there to even exist a Lipschitz constant that would force an interval to be selected for division in this way, that interval must have the smallest midpoint value of all intervals having the same length. Moreover, there should be no interval of a different length for which $f(c) - L(b-a)/2$ was smaller.

The DIRECT algorithm applies this rule to all possible combinations of possible Lipschitz constants and interval sizes. If one plots the values of f at the midpoints against the lengths of the intervals in the list to obtain a plot like the one in Figure 8.10, one can visually eliminate all but one interval for each interval length. By taking the convex hull of the lowest points, one can eliminate interval lengths for which all function values are so high that $f(c) - L(b-a)/2$ would be smaller for the best point at a different length no matter what L was. For example, the three points that intersect the line in Figure 8.10 would correspond to intervals that would be subdivided at this step. The slopes of the line segments through the three points are estimates of the Lipschitz constant. These estimates are not used explicitly, as they would be in the Shubert algorithm, but implicitly in the process of selection of intervals to be divided. Unlike the Shubert algorithm, where the Lipschitz constant is assumed known, the DIRECT algorithm will eventually subdivide every interval.

The resulting algorithm may divide more than a single interval at each stage and the number of intervals to be divided may vary. This is easy to implement for a single variable. However, for more than one variable there are several ways to divide a hyperrectangle into parts and one must keep track of how an interval has previously been divided in order not to cluster sample points prematurely by repeatedly dividing an interval in the same way. Figures 8.11 and 8.12, taken from [108], illustrate this issue for $N = 2$. In Figure 8.11 the entire rectangle will be divided. Shading indicates that the rectangle has been selected for division. Four new midpoints are sampled. The subdivision into new rectangles could be done in two ways: the figure shows an initial horizontal subdivision followed by a vertical division of the rectangle that contains the original center. The second division is shown in Figure 8.12. The two shaded rectangles are selected for division. Note that four new centers are added to the small square and two to the larger, nonsquare, rectangle. In this way the minimum number of new centers is added.

DIRECT parallelizes in a natural way. All hyperrectangles that are candidates for division may be divided simultaneously, and for each hyperrectangle the function evaluations at each of the new midpoints can also be done in parallel. We refer the reader to [150] and [108] for details on the data structures and to [108] for a FORTRAN implementation and additional discussion on the exploitation of parallelism.

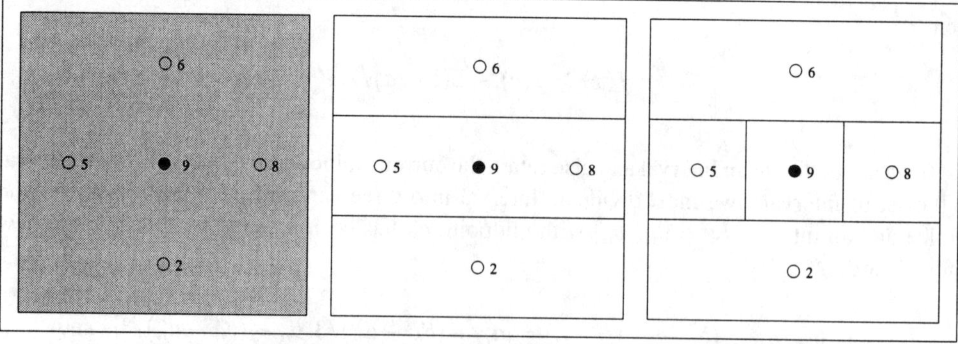

Figure 8.11: *Initial Division of Rectangles with DIRECT*

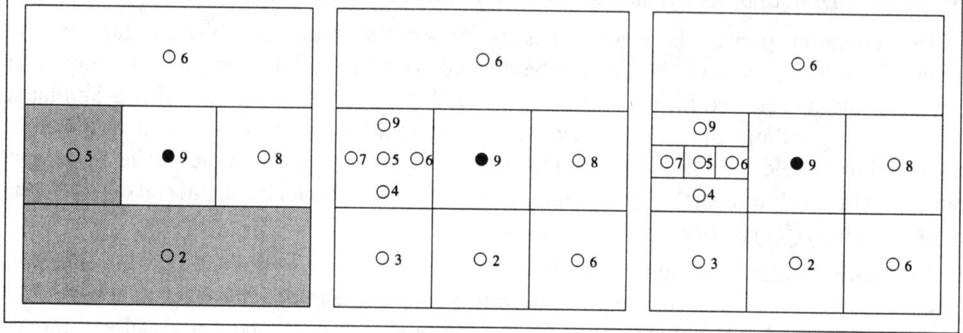

Figure 8.12: *Second Division of Rectangles with DIRECT*

8.5 Examples

In each of the examples we compare the central difference BFGS form of implicit filtering from §7.6 (solid line) with the Nelder–Mead (dashed line), Hooke–Jeeves (solid line with circles), and MDS (dashed-dotted line) algorithms.

For each example we specified both an initial iterate and choice of scales. This is sufficient to initialize both implicit filtering and Hooke–Jeeves. We used the implicit filtering forward difference stencil as the initial simplex for both Nelder–Mead and MDS.

The plots reflect the differences in the startup procedures for the varying algorithms. In particular, Nelder–Mead and MDS sort the simplex and hence, if the initial iterate is not the best point, report the lower value as the first iterate.

The relative performance of the various methods on these example problems should not be taken as a definitive evaluation, nor should these examples be thought of as a complete suite of test problems. One very significant factor that is not reflected in the results in this section is that both implicit filtering [69], [55] and multidirectional search [85] are easy to implement in parallel, while Nelder–Mead and Hooke–Jeeves are inherently sequential. The natural parallelism of implicit filtering and multidirectional search can be further exploited by using idle processors to explore other points on the line search direction or the pattern.

8.5.1 Weber's Problem

The initial data were the same as that in §7.6.1. Implicit filtering does relatively poorly for this problem because of the nonsmoothness at optimality. The resluts for these problems are plotted

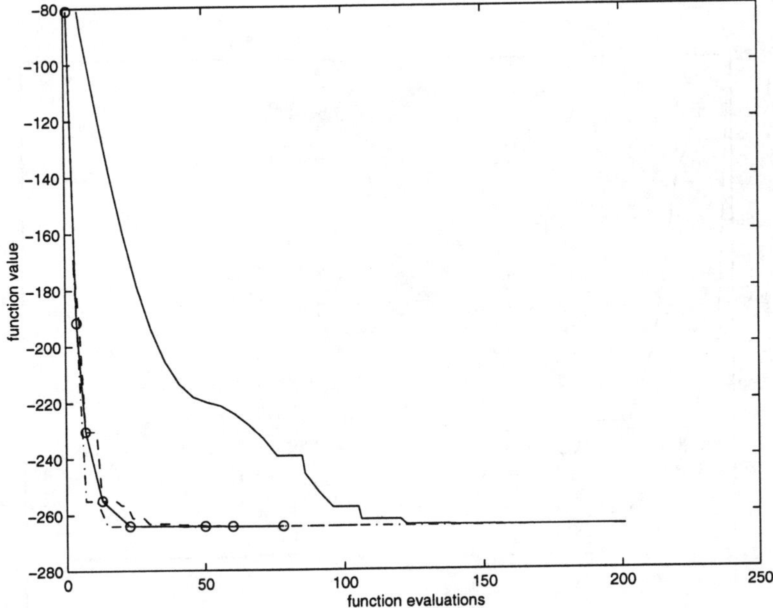

Figure 8.13: *First Weber Example*

in Figures 8.13, 8.14, and 8.15. The other three algorithms perform equally well. Note in the third example that MDS finds a local minimum that is not the global minimum.

8.5.2 Parameter ID

In the computations reported in this section each algorithm was allowed 500 evaluations of f and the sequence of scales was $\{2^{-j}\}_{j=1}^{12}$.

We begin with the two examples from §7.6.2. With the initial iterate of $(5,5)^T$, the exact solution to the continuous problem lies on the grid that the Hooke–Jeeves algorithm uses to search for the solution. This explains the unusually good performance of the Hooke–Jeeves optimization shown in both Figures 8.16 and 8.17. When the initial iterate is changed to $(5.1, 5.3)^T$, the performance of Hooke–Jeeves is very different as one can see from Figures 8.18 and 8.19. The other algorithms do not have such a sensitivity to the initial iterate for this example. We have no explanation for the good performance turned in by the Nelder–Mead algorithm on this problem.

8.5.3 Convex Quadratics

The problems and initial data are the same as those in §7.6.3. This is an example of how sampling algorithms can perform poorly for very simple problems and how this poor performance is made worse by increasing the problem size. Exercise 7.7.4 illustrates this point very directly. One would expect implicit filtering to do well since a central difference gradient has no error for quadratic problems. For the larger problem ($N = 32$, Figures 8.21 and 8.23), both the Nelder–Mead and MDS algorithms perform poorly while the Hooke–Jeeves algorithm does surprisingly well. The difference in performance of the algorithms is much smaller for the low-dimensional problem ($N = 4$, Figures 8.20 and 8.22).

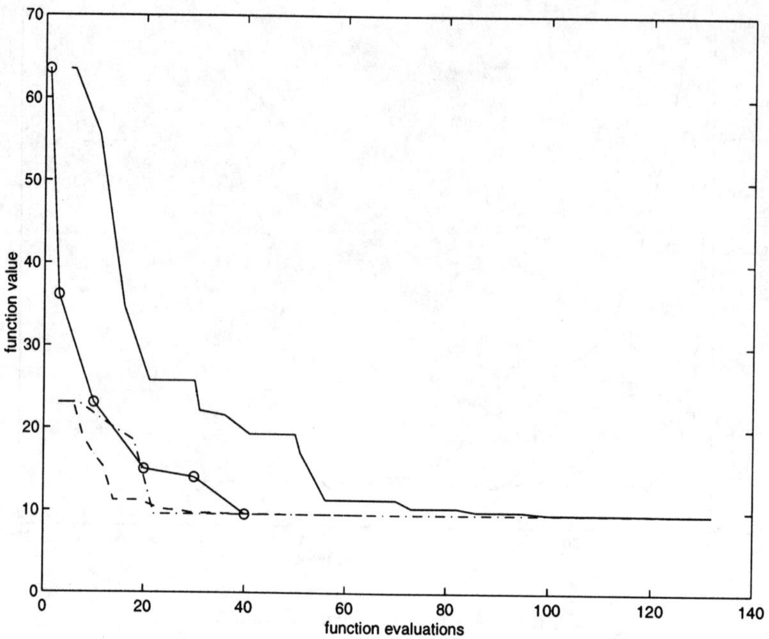

Figure 8.14: *Second Weber Example*

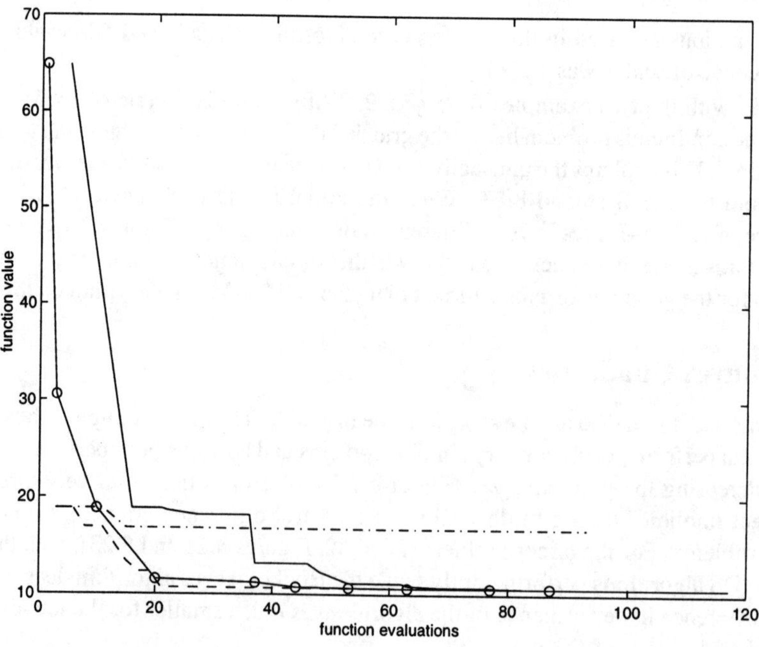

Figure 8.15: *Third Weber Example*

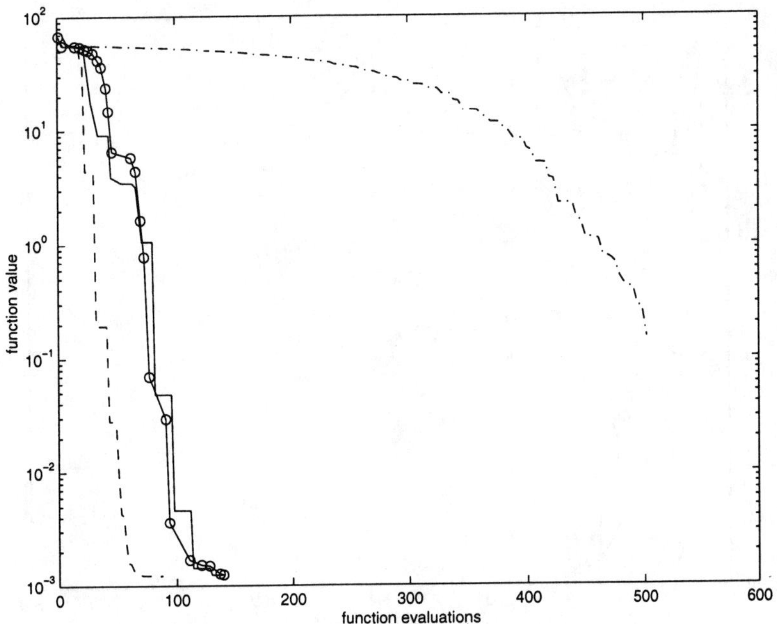

Figure 8.16: *Parameter ID*, $tol = 10^{-3}$, $x_0 = (5,5)^T$

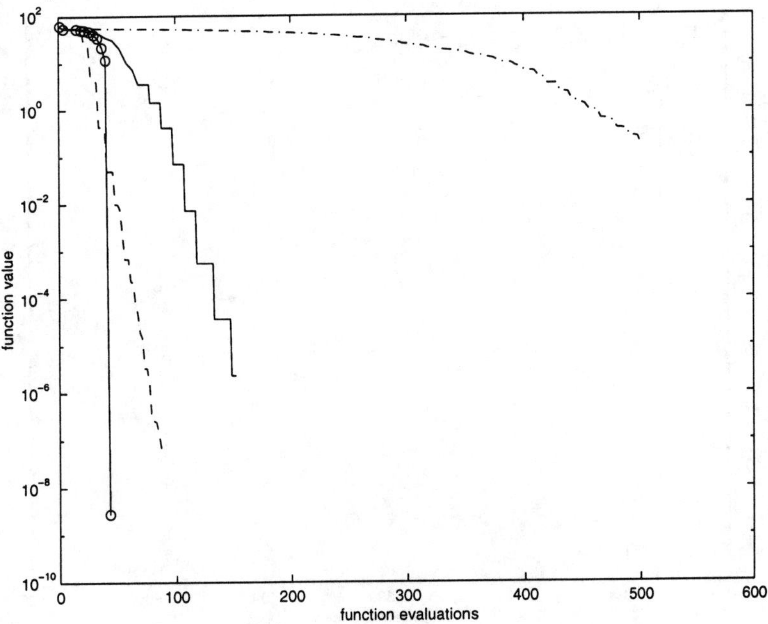

Figure 8.17: *Parameter ID*, $tol = 10^{-6}$, $x_0 = (5,5)^T$

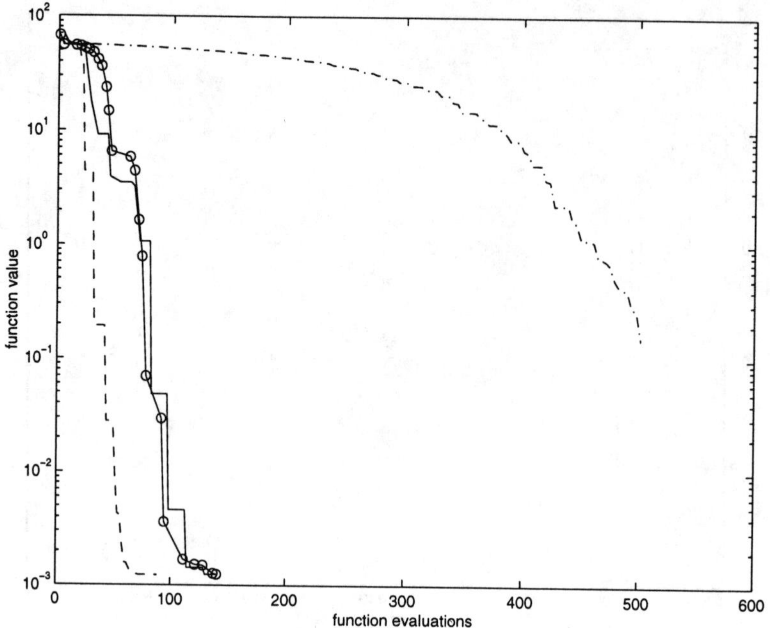

Figure 8.18: *Parameter ID*, $tol = 10^{-3}$, $x_0 = (5.1, 5.3)^T$

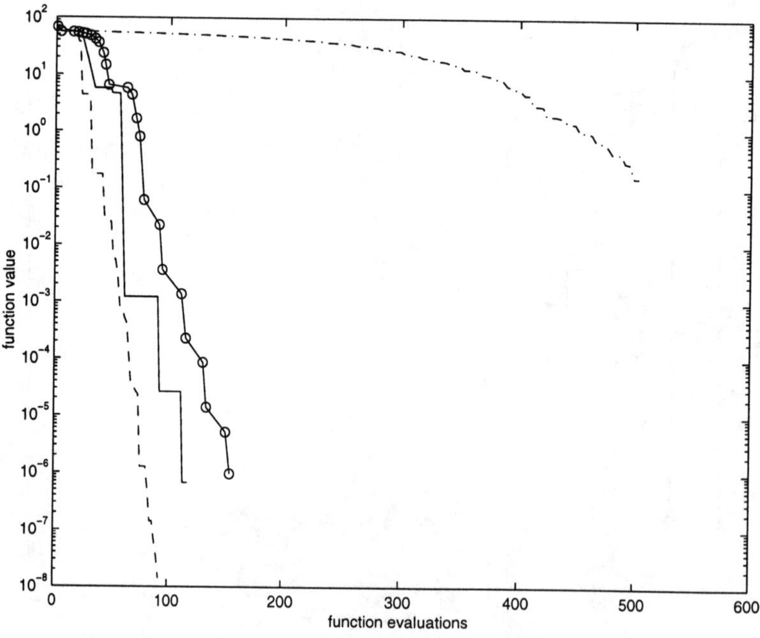

Figure 8.19: *Parameter ID*, $tol = 10^{-6}$, $x_0 = (5.1, 5.3)^T$

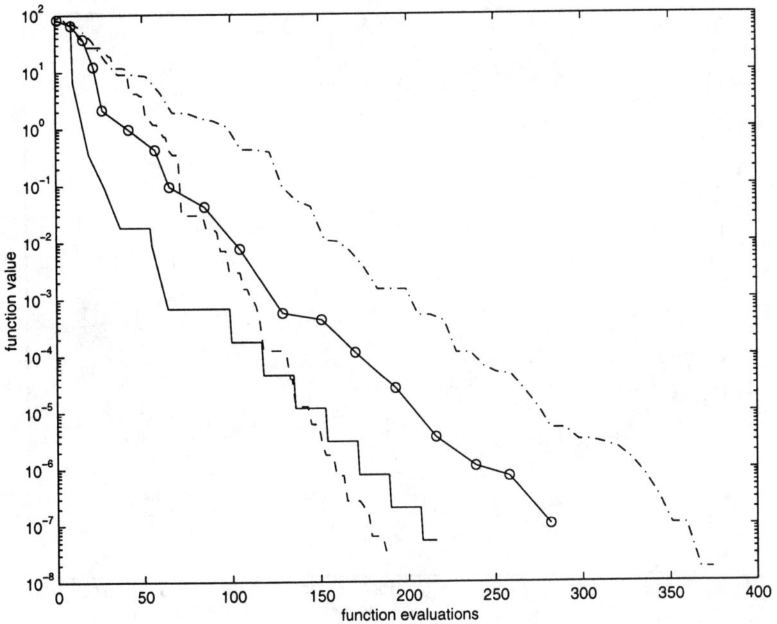

Figure 8.20: *Unperturbed Quadratic,* $N = 4$

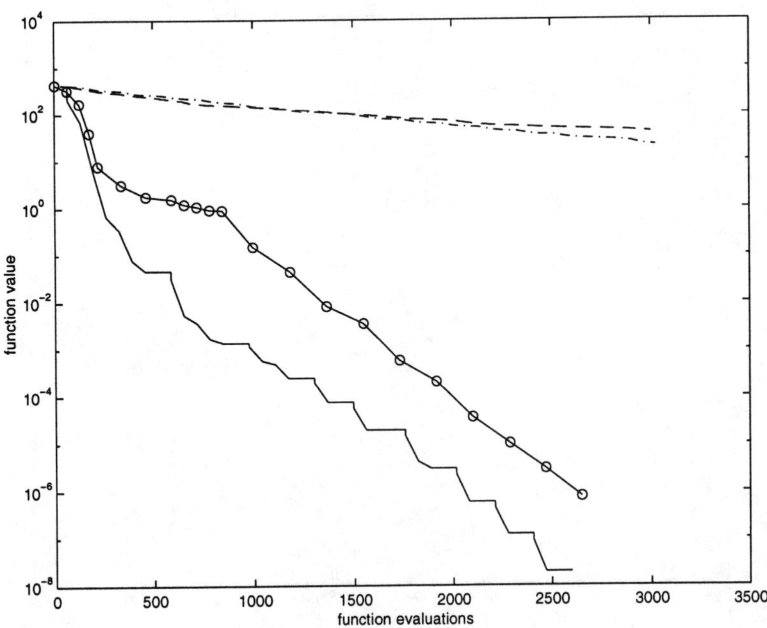

Figure 8.21: *Unperturbed Quadratic,* $N = 32$

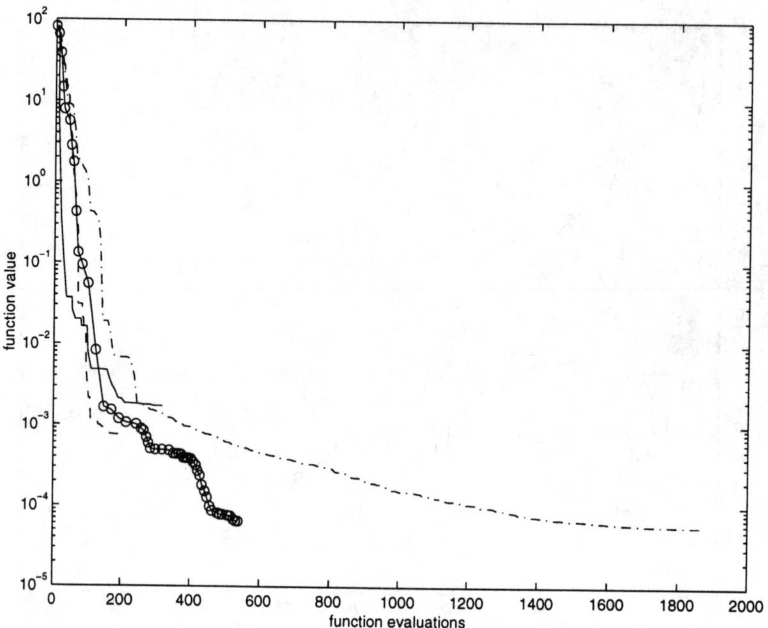

Figure 8.22: *Perturbed Quadratic,* $N = 4$

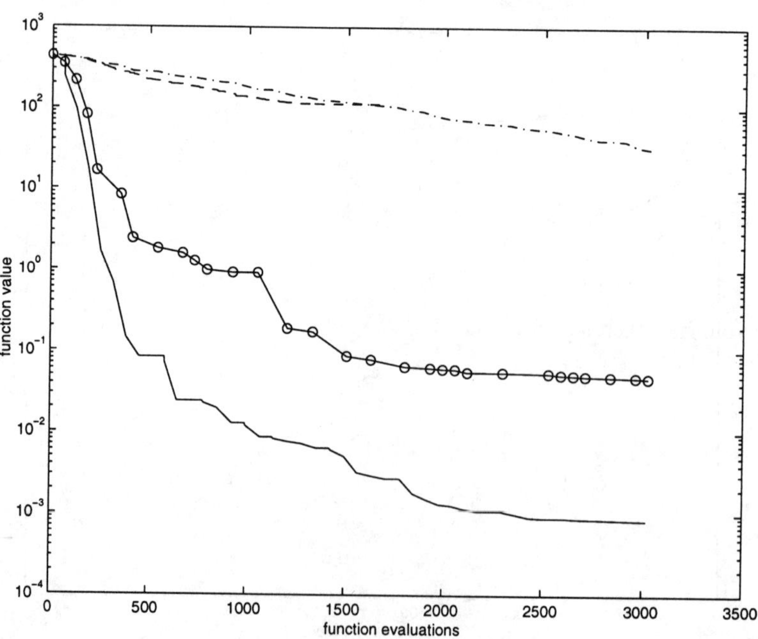

Figure 8.23: *Perturbed Quadratic,* $N = 32$

8.6 Exercises on Search Algorithms

8.6.1. Let S_l for $1 \le l \le 3$ be the simplices having one vertex at $(x^l)_1 = (10, 10, 10, 10)$ and direction vectors V^l given by

$$V^1 = \text{diag}(1, 2, 3, 4), \text{V}^2 = \text{diag}(4, 3, 2, 1), \text{V}^3 = \text{diag}(2, 2, 2, 2).$$

For each $l = 1, 2, 3$, apply the Nelder–Mead algorithm to the function f defined for $x \in R^4$ by

$$(x_1 - x_2 x_3 x_4)^2 + (x_2 - x_3 x_4)^2 + (x_3 - x_4)^2 + x_4^2$$

with the initial simplex V^l. What happened? This example is one of Nelder's favorites [203].

8.6.2. Show that if the set $\{x \mid f(x) \le f(x_1^0)\}$ is bounded and S_0 is either an equilateral or a right simplex, then (8.9) holds.

8.6.3. One can modify MDS [263] by eliminating the expansion step and only computing reflected points until one is found that is better than x_1. If no reflected points are better, then perform a contraction step. Prove that Theorem 8.2.1 holds for this implementation. Implement MDS in this way and compare it with Algorithm `mds`. Are the savings in calls to f for each iterate realized in a savings for the entire optimization?

8.6.4. The easiest problem in optimization is to minimize $x^T x$. Give the algorithms in this section a chance to show what they can do by using them to solve this problem. Try several initial iterates (or initial simplices/patterns) and several problem dimensions (especially $N = 8, 16, 32$).

8.6.5. The search methods in this section impose a structure on the sampling and thereby hope to find a useful optimal point far more efficiently than using an unstructured deterministic or random search. Implement an unstructured search and use your algorithm to minimize $x^T x$ when $N = 2$. For an example of such a method, take the one from [6], please.

8.6.6. The *Spendley, Hext, and Himsworth algorithm* [244] manages the simplices in a very different way from those we've discussed in the text. Use the information in [244] and [267] to implement this algorithm. Use Theorem 6.2.9 to prove convergence for $N = 2$. What happens to both your implementation and analysis when you try $N = 3$ or arbitrary N? Explain Table 5 in [244].

8.6.7. Use any means necessary to solve the Lennard–Jones problem. Have your results improved since you tried exercises 6.4.3 and 7.7.4?

Bibliography

[1] E. Aarts and J. Korst, *Simulated annealing and Boltzmann machines*, Wiley, New York, 1989.

[2] L. Adams and J. L. Nazareth, eds., *Linear and Nonlinear Conjugate Gradient Methods*, SIAM, Philadelphia, 1996.

[3] M. Al-Baali, *Descent property and global convergence of the Fletcher-Reeves method with inexact line searches*, IMA J. Numer. Anal., 5 (1985), pp. 121–124.

[4] N. Alexandrov, J. E. Dennis, R. M. Lewis, and V. Torczon, *A trust region framework for managing the use of approximation models in optimization*, Structural Optimization, 15 (1998), pp. 16–23.

[5] E. L. Allgower and K. Georg, *Numerical path following*, in Handbook of Numerical Analysis, P. G. Ciarlet and J. L. Lions, eds., vol. 5, North Holland, 1997, pp. 3–207.

[6] Anonymous, *A new algorithm for optimization*, Math. Prog., 3 (1972), pp. 124–128.

[7] L. Armijo, *Minimization of functions having Lipschitz-continuous first partial derivatives*, Pacific J. Math., 16 (1966), pp. 1–3.

[8] U. M. Ascher and L. R. Petzold, *Computer Methods for Ordinary Differential Equations and Differential Algebraic Equations*, SIAM, Philadelphia, 1998.

[9] K. E. Atkinson, *Iterative variants of the Nyström method for the numerical solution of integral equations*, Numer. Math., 22 (1973), pp. 17–31.

[10] B. M. Averick and J. J. Moré, *User guide for the MINPACK-2 test problem collection*, Tech. Rep. ANL/MCS-TM-157, Argonne National Laboratory, Math. and Comp. Science Div. Report, October 1991.

[11] O. Axelsson, *Iterative Solution Methods*, Cambridge University Press, Cambridge, 1994.

[12] S. Banach, *Sur les opérations dans les ensembles abstraits et leur applications aux équations intégrales*, Fund. Math, 3 (1922), pp. 133–181.

[13] H. T. Banks and H. T. Tran, *Mathematical and experimental modeling of physical processes*. Department of Mathematics, North Carolina State University, unpublished lecture notes for Mathematics 573-4, 1997.

[14] M. S. Barlett, *An inverse matrix adjustment arising in discriminant analysis*, Ann. Math. Stat., 22 (1951), pp. 107–111.

[15] R. Barrett, M. Berry, T. F. Chan, J. Demmel, J. Donato, J. Dongarra, V. Eijkhout, R. Pozo, C. Romine, and H. van der Vorst, *Templates for the Solution of Linear Systems: Building Blocks for Iterative Methods*, SIAM, Philadelphia, 1993.

[16] K. J. Bathe and A. P. Cimento, *Some practical procedures for the solution of nonlinear finite element equations*, Comp. Meth. Appl. Mech. Eng., 22 (1980), pp. 59–85.

[17] A. Ben-Israel, *A Newton-Raphson method for the solution of systems of equations*, J. Math. Anal. Appl., 15 (1966), pp. 243–252.

[18] D. P. Bertsekas, *On the Goldstein-Levitin-Polyak gradient projection method*, IEEE Trans. Autom. Control, 21 (1976), pp. 174–184.

[19] ——, *Projected Newton methods for optimization problems with simple constraints*, SIAM J. Control Optim., 20 (1982), pp. 221–246.

[20] J. T. Betts, *An improved penalty function method for solving constrained parameter optimization problems*, J. Optim. Theory Appl., 16 (1975), pp. 1–24.

[21] ——, *Solving the nonlinear least square problem: application of a general method*, J. Optim. Theory Appl., 18 (1976), pp. 469–483.

[22] J. T. Betts, M. J. Carter, and W. P. Huffman, *Software for nonlinear optimization*, Tech. Rep. MEA-LR-083 R1, Mathematics and Engineering Analysis Library Report, Boeing Information and Support Services, June 6 1997.

[23] J. T. Betts and P. D. Frank, *A sparse nonlinear optimization algorithm*, J. Optim. Theory Appl., 82 (1994), pp. 519–541.

[24] P. T. Boggs, *The convergence of the Ben-Israel iteration for nonlinear least squares problems*, Math. Comp., 30 (1976), pp. 512–522.

[25] P. T. Boggs and J. E. Dennis, *A stability analysis for perturbed nonlinear iterative methods*, Math. Comp., 30 (1976), pp. 1–17.

[26] I. Bongatz, A. R. Conn, and P. L. Toint, *CUTE: Constrained and Unconstrained Testing Environment*, ACM transactions on mathematical software, 21 (1995), pp. 123–160.

[27] A. J. Booker, *DOE for computer output*, Tech. Rep. BCSTECH-94-052, Boeing Computer Services, Seattle, WA, 1994.

[28] A. J. Booker, J. E. Dennis, P. D. Frank, D. B. Serafini, V. Torczon, and M. W. Trosset, *A rigorous framework for optimization of expensive function by surrogates*, Structural Optimization, 17 (1999), pp. 1–13.

[29] D. M. Bortz and C. T. Kelley, *The simplex gradient and noisy optimization problems*, in Computational Methods in Optimal Design and Control, J. T. Borggaard, J. Burns, E. Cliff, and S. Schreck, eds., vol. 24 of Progress in Systems and Control Theory, Birkhäuser, Boston, 1998, pp. 77–90.

[30] A. Bouaricha, *Tensor methods for large, sparse, unconstrained optimization*, SIAM J. Optim., 7 (1997), pp. 732–756.

[31] H. Brakhage, *Über die numerische Behandlung von Integralgleichungen nach der Quadraturformelmethode*, Numer. Math., 2 (1960), pp. 183–196.

[32] K. E. Brenan, S. L. Campbell, and L. R. Petzold, *The Numerical Solution of Initial Value Problems in Differential-Algebraic Equations*, no. 14 in Classics in Applied Mathematics, SIAM, Philadelphia, 1996.

[33] P. N. Brown and Y. Saad, *Convergence theory of nonlinear Newton-Krylov algorithms*, SIAM J. Optim., 4 (1994), pp. 297–330.

[34] C. G. Broyden, *A class of methods for solving nonlinear simultaneous equations*, Math. Comp., 19 (1965), pp. 577–593.

[35] ——, *Quasi-Newton methods and their application to function minimization*, Math. Comp., 21 (1967), pp. 368–381.

[36] ——, *A new double-rank minimization algorithm*, AMS Notices, 16 (1969), p. 670.

[37] C. G. Broyden, J. E. Dennis, and J. J. Moré, *On the local and superlinear convergence of quasi-Newton methods*, J. Inst. Maths. Applics., 12 (1973), pp. 223–246.

[38] R. H. Byrd, T. Derby, E. Eskow, K. P. B. Oldenkamp, and R. B. Schnabel, *A new stochastic/perturbation method for large-scale global optimization and its application to water cluster problems*, in Large Scale Optimization: State of the Art, W. W. Hager, D. W. Hearn, and P. Pardalos, eds., Boston, 1994, Kluwer Academic Publishers B.V., pp. 68–81.

[39] R. H. Byrd, C. L. Dert, A. H. G. R. Kan, and R. B. Schnabel, *Concurrent stochastic methods for global optimization*, Math. Prog., 46 (1990), pp. 1–30.

[40] R. H. Byrd, E. Eskow, and R. B. Schnabel, *A new large-scale global optimization method and its application to Lennard-Jones problems*, Tech. Rep. CU-CS-630-92, University of Colorado at Boulder, November 1992.

[41] R. H. Byrd, H. F. Khalfan, and R. B. Schnabel, *Analysis of a symmetric rank-one trust region method*, SIAM J. Optim., 6 (1996), pp. 1025–1039.

[42] R. H. Byrd, P. Lu, J. Nocedal, and C. Zhu, *A limited memory algorithm for bound constrained optimization*, SIAM J. Sci. Statist. Comput., 16 (1995), pp. 1190–1208.

[43] R. H. Byrd and J. Nocedal, *A tool for the analysis of quasi-Newton methods with application to unconstrained minimization*, SIAM J. Numer. Anal., 26 (1989), pp. 727–739.

[44] R. H. Byrd, J. Nocedal, and R. B. Schnabel, *Representation of quasi-Newton matrices and their use in limited memory methods*, Mathematical Programming, 63 (1994), pp. 129–156.

[45] R. H. Byrd, J. Nocedal, and Y. Yuan, *Global convergence of a class of quasi-Newton methods on convex problems*, SIAM J. Numer. Anal., 24 (1987), pp. 1171–1190.

[46] R. H. Byrd, R. B. Schnabel, and G. A. Schultz, *Parallel quasi-Newton methods for unconstrained optimization*, Math. Prog., 42 (1988), pp. 273–306.

[47] P. H. Calamai and J. Moré, *Projected gradient methods for linearly constrained problems*, Mathematical Programming, 39 (1987), pp. 93–116.

[48] S. L. Campbell, C. T. Kelley, and K. D. Yeomans, *Consistent initial conditions for unstructured higher index DAEs: A computational study*, in Proceedings of Conference on Computational Engineering in Systems Applications (CESA'96), Lille, France, 1996, pp. 416–421.

[49] S. L. CAMPBELL AND C. D. MEYER, *Generalized Inverses of Linear Transformations*, Dover Press, New York, 1991.

[50] S. L. CAMPBELL AND K. D. YEOMANS, *Behavior of the nonunique terms in general DAE integrators*, Appl. Num. Math., 28 (1998), pp. 209–226.

[51] R. G. CARTER, *On the global convergence of trust region algorithms using inexact gradient information*, SIAM J. Numer. Anal., 28 (1991), pp. 251–265.

[52] A. CAUCHY, *Methode generale pour la resolution des systemes d'equations simultanees*, Comp. Rend. Acad. Sci. Paris, (1847), pp. 536–538.

[53] T. D. CHOI, 1998. Private Communication.

[54] ——, *Bound Constrained Optimization*, PhD thesis, North Carolina State University, Raleigh, North Carolina, 1999.

[55] T. D. CHOI, O. J. ESLINGER, C. T. KELLEY, J. W. DAVID, AND M. ETHERIDGE, *Optimization of automotive valve train components with implicit filtering*, Optim. Engrg., 1 (2000), pp. 9–28.

[56] T. D. CHOI AND C. T. KELLEY, *Superlinear convergence and implicit filtering*, SIAM J. Optim., 10 (2000), pp. 1149–1162.

[57] T. F. COLEMAN AND Y. LI, *On the convergence of interior-reflective Newton methods for nonlinear minimization subject to bounds*, Math. Prog., 67 (1994), pp. 189–224.

[58] ——, *An interior trust region approach for nonlinear minimization subject to bounds*, SIAM J. Optim., 6 (1996), pp. 418–445.

[59] T. F. COLEMAN AND J. J. MORÉ, *Estimation of sparse Jacobian matrices and graph coloring problems*, SIAM J. Numer. Anal., 20 (1983), pp. 187–209.

[60] P. CONCUS, G. H. GOLUB, AND D. P. O'LEARY, *A generalized conjugate gradient method for the numerical solution of elliptic partial differential equations*, in Sparse Matrix Computations, J. R. Bunch and D. J. Rose, eds., Academic Press, 1976, pp. 309–332.

[61] A. R. CONN, N. I. M. GOULD, AND P. L. TOINT, *Global convergence of a class of trust region algorithms for optimization problems with simple bounds*, SIAM J. Numer. Anal., 25 (1988), pp. 433–460.

[62] ——, *Testing a class of methods for solving minimization problems with simple bounds on the variables*, Math. Comp., 50 (1988), pp. 399–430.

[63] ——, *Convergence of quasi-Newton matrices generated by the symmetric rank one update*, Math. Programming A, 50 (1991), pp. 177–195.

[64] ——, *LANCELOT: A Fortran Package for Large-Scale Nonlinear Optimization (Release A)*, no. 17 in Springer Series in Computational Mathematics, Springer Verlag, Heidelberg, Berlin, New York, 1992.

[65] A. R. CONN, K. SCHEINBERG, AND P. L. TOINT, *On the convergence of derivative-free methods for unconstrained optimization*, in Approximation Theory and Optimization: Tributes to M. J. D. Powell, A. Iserles and M. Buhmann, eds., Cambridge, U.K., 1997, Cambridge University Press, pp. 83–108.

[66] ——, *Recent progress in unconstrained optimization without derivatives*, Math. Prog. Ser. B, 79 (1997), pp. 397–414.

[67] A. R. CURTIS, M. J. D. POWELL, AND J. K. REID, *On the estimation of sparse Jacobian matrices*, J. Inst. Math. Appl., 13 (1974), pp. 117–119.

[68] J. W. DANIEL, *The conjugate gradient method for linear and nonlinear operator equations*, SIAM J. Numer. Anal., 4 (1967), pp. 10–26.

[69] J. W. DAVID, C. Y. CHENG, T. D. CHOI, C. T. KELLEY, AND J. GABLONSKY, *Optimal design of high speed mechanical systems*, Tech. Rep. CRSC-TR97-18, North Carolina State University, Center for Research in Scientific Computation, July 1997. Mathematical Modeling and Scientific Computing, to appear in Vol 9.

[70] J. W. DAVID, C. T. KELLEY, AND C. Y. CHENG, *Use of an implicit filtering algorithm for mechanical system parameter identification*. SAE Paper 960358, year=1996, 1996 SAE International Congress and Exposition Conference Proceedings, Modeling of CI and SI Engines, pp. 189–194, Society of Automotive Engineers, Washington, DC.

[71] W. C. DAVIDON, *Variable metric methods for minimization*, Tech. Rep. ANL-5990, Argonne National Laboratory, 1959.

[72] ——, *Variable metric methods for minimization*, SIAM J. Optim., 1 (1991), pp. 1–17.

[73] T. J. DEKKER, *Finding a zero by means of successive linear interpolation*, in Constructive Aspects of the Fundamental Theorem of Algebra, B. Dejon and P. Henrici, eds., New York, 1969, Wiley-Interscience, pp. 37–48.

[74] R. DEMBO, S. EISENSTAT, AND T. STEIHAUG, *Inexact Newton methods*, SIAM J. Numer. Anal., 19 (1982), pp. 400–408.

[75] R. DEMBO AND T. STEIHAUG, *Truncated Newton algorithms for large-scale optimization*, Math. Prog., 26 (1983), pp. 190–212.

[76] J. E. DENNIS, *Nonlinear least squares and equations*, in The State of the Art in Numerical Analysis, D. Jacobs, ed., London, 1977, Academic Press, pp. 269–312.

[77] J. E. DENNIS, D. M. GAY, AND R. E.WELSCH, *An adaptive nonlinear least-squares algorithm*, ACM Trans. Math. Software, 7 (1981), pp. 348–368.

[78] ——, *Algorithm 573: NL2SOL – An adaptive nonlinear least-squares algorithm*, ACM Trans. Math. Software, 7 (1981), pp. 369–383.

[79] J. E. DENNIS, M. HEINKENSCHLOSS, AND L. N. VICENTE, *Trust-region interior-point algorithms for a class of nonlinear programming problems*, SIAM J. Control and Optimization, 36 (1998), pp. 1750–1794.

[80] J. E. DENNIS AND H. H. W. MEI, *Two unconstrained optimization algorithms which use function and gradient values*, J. Optim. Theory Appl., 28 (1979), pp. 453–482.

[81] J. E. DENNIS AND J. J. MORÉ, *A characterization of superlinear convergence and its application to quasi-Newton methods*, Math. Comp, 28 (1974), pp. 549–560.

[82] ——, *Quasi-Newton methods, methods, motivation and theory*, SIAM Review, 19 (1977), pp. 46–89.

[83] J. E. DENNIS AND R. B. SCHNABEL, *Least change secant updates for quasi-Newton methods*, SIAM Review, 21 (1979), pp. 443–459.

[84] ——, *Numerical Methods for Unconstrained Optimization and Nonlinear Equations*, no. 16 in Classics in Applied Mathematics, SIAM, Philadelphia, 1996.

[85] J. E. DENNIS AND V. TORCZON, *Direct search methods on parallel machines*, SIAM J. Optim., 1 (1991), pp. 448 – 474.

[86] J. E. DENNIS AND L. N. VICENTE, *Trust-region interior-point algorithms for minimization problems with simple bounds*, in Applied Mathematics and Parallel Computing, H. Fischer, B. Riedmiller, and S. Schaffler, eds., Hidelberg, 1997, Springer, pp. 97–109.

[87] J. E. DENNIS AND H. F. WALKER, *Convergence theorems for least change secant update methods*, SIAM J. Numer. Anal., 18 (1981), pp. 949–987.

[88] ——, *Inaccuracy in quasi-Newton methods: Local improvement theorems*, in Mathematical Programming Study 22: Mathematical programming at Oberwolfach II, North–Holland, Amsterdam, 1984, pp. 70–85.

[89] J. E. DENNIS AND D. J. WOODS, *Optimization on microcomputers: the Nelder-Mead simplex algorithm*, in New Computing Environments: Microcomputers in Large-Scale Computing, A. Wouk, ed., Philadelphia, 1987, SIAM, pp. 116–122.

[90] P. DEUFLHARD AND V. APOSTOLESCU, *A study of the Gauss-Newton algorithm for the soution of nonlinear least squares problems*, Tech. Rep. 51, Univ. Heidelberg Preprint, 1980.

[91] P. DEUFLHARD, R. W. FREUND, AND A. WALTER, *Fast secant methods for the iterative solution of large nonsymmetric linear systems*, Impact of Computing in Science and Engineering, 2 (1990), pp. 244–276.

[92] P. DEUFLHARD AND G. HEINDL, *Affine invariant convergence theorems for Newton's method and extensions to related methods*, SIAM J. Numer. Anal., 16 (1979), pp. 1–10.

[93] W. J. DUNCAN, *Some devices for the solution of large sets of simultaneous linear equations (with an appendix on the reciprocation of partitioned matrices)*, The London, Edinburgh, and Dublin Philosophical Magazine and Journal of Science, Seventh Series, 35 (1944), pp. 660–670.

[94] J. C. DUNN, *Global and asymptotic convergence rate estimates for a class of projected gradient processes*, SIAM J. Control Optim., 19 (1981), pp. 368–400.

[95] ——, *On the convergence of projected gradient processes to singular critical points*, J. Optim. Th. Appl., 55 (1987), pp. 203–215.

[96] ——, *A projected Newton method for minimization problems with nonlinear inequality constraints*, Numer. Math., 53 (1988), pp. 377–409.

[97] J. C. DUNN AND E. W. SACHS, *The effect of perturbations on the convergence rates of optimization algorithms*, Applied Math. and Optimization, 10 (1983), pp. 143–147.

[98] J. C. DUNN AND T. TIAN, *Variants of the Kuhn-Tucker sufficient conditions in cones of nonnegative functions*, SIAM J. Control and Optimization, 30 (1992), pp. 1361–1384.

[99] S. C. EISENSTAT AND H. F. WALKER, *Choosing the forcing terms in an inexact Newton method*, SIAM J. Sci. Comput., 17 (1996), pp. 16–32.

[100] E. ESKOW AND R. B. SCHNABEL, *Algorithm 695: Software for a new modified Cholesky factorization*, ACM Transactions on Mathematical Software, 17 (1991), pp. 306–312.

[101] A. V. FIACCO AND G. P. MCCORMICK, *Nonlinear Programming*, no. 4 in Classics in Applied Mathematics, SIAM, Philadelphia, 1990.

[102] R. FLETCHER, *Generalized inverse methods for the best least squares solution of systems of nonlinear equations*, Computer J., 10 (1968), pp. 392–399.

[103] ——, *A new approach to variable metric methods*, Comput. J., 13 (1970), pp. 317–322.

[104] ——, *Practical methods of optimization*, John Wiley and Sons, New York, 1987.

[105] R. FLETCHER AND M. J. D. POWELL, *A rapidly convergent descent method for minimization*, Computer J., 6 (1963), pp. 163–168.

[106] R. FLETCHER AND C. M. REEVES, *Function minimization by conjugate gradients*, Comp. J., (1964), pp. 149–154.

[107] S. J. FORTUNE, D. M. GAY, B. W. KERNIGHAN, O. LANDRON, R. A. VALENZUELA, AND M. H. WRIGHT, *WISE design of indoor wireless systems*, IEEE Computational Science and Engineering, 2 (1995), pp. 58–68.

[108] J. GABLONSKY, *An implementation of the DIRECT algorithm*, Tech. Rep. CRSC-TR98-29, North Carolina State University, Center for Research in Scientific Computation, August 1998.

[109] D. M. GAY, *Computing localy constrained steps*, SIAM J. Sci. Statist. Comput., 2 (1981).

[110] C. W. GEAR, *Numerical Initial Value Problems in Ordinary Differential Equations*, Prentice-Hall, Englewood Cliffs, 1971.

[111] R. R. GERBER AND F. T. LUK, *A generalized Broyden's method for solving simultaneous linear equations*, SIAM J. Numer. Anal., 18 (1981), pp. 882–890.

[112] J. C. GILBERT AND J. NOCEDAL, *Global convergence properties of conjugate gradient methods for optimization*, SIAM J. Optim., (1992), pp. 21–42.

[113] P. E. GILL AND W. MURRAY, *Newton-type methods for unconstrained and linearly constrained optimization*, Math. Prog., 28 (1974), pp. 311–350.

[114] ——, *Safeguarded steplength algorithms for optimization using descent methods*, Tech. Rep. NAC 37, National Physical Laboratory Report, Teddington, England, 1974.

[115] ——, *Non-linear least squares and nonlinearly constrained optimization*, in Numerical Analysis, Dundee, 1975, vol. 506 of Springer-Verlag Lecture Notes in Mathematics, Springer-Verlag, Berlin, 1976.

[116] ——, *Algorithms for the solution of nonlinear least squares problems*, SIAM J. Numer. Anal., 15 (1978), pp. 977–992.

[117] P. E. GILL, W. MURRAY, AND M. H. WRIGHT, *Practical Optimization*, Academic Press, London, 1981.

[118] P. GILMORE, *An Algorithm for Optimizing Functions with Multiple Minima*, PhD thesis, North Carolina State University, Raleigh, North Carolina, 1993.

[119] ——, *IFFCO: Implicit Filtering for Constrained Optimization*, Tech. Rep. CRSC-TR93-7, Center for Research in Scientific Computation, North Carolina State University, May 1993. available by anonymous ftp from ftp.math.ncsu.edu in FTP/kelley/iffco/ug.ps.

[120] P. Gilmore and C. T. Kelley, *An implicit filtering algorithm for optimization of functions with many local minima*, SIAM J. Optim., 5 (1995), pp. 269–285.

[121] P. A. Gilmore, S. S. Berger, R. F. Burr, and J. A. Burns, *Automated optimization techniques for phase change piezoelectric ink jet performance enhancement*, in 1997 International Conference on Digital Printing Technologies, Society for Imaging Science and Technology, IS&T's NIP 13, November, 1997, pp. 716–721.

[122] R. Glowinski, *Numerical Methods for Nonlinear Variational Problems*, Springer-Verlag, New York, 1984.

[123] D. E. Goldberg, *Genetic Algorithms in Search, Optimization, and Machine Learning*, Addison Wesely, Reading, Mass., 1989.

[124] D. Goldfarb, *A family of variable metric methods derived by variational means*, Math. Comp., 24 (1970), pp. 23–26.

[125] A. A. Goldstein, *Constructive Real Analysis*, Harper and Row, New York, 1967.

[126] G. H. Golub and V. Pereyra, *The differentiation of pseudo-inverses and nonlinear least squares problems whose variables separate*, SIAM J. Numer. Anal., 10 (1973), pp. 413–432.

[127] G. H. Golub and C. G. VanLoan, *Matrix Computations*, Johns Hopkins University Press, Baltimore, 1983.

[128] A. Greenbaum, *Iterative Methods for Solving Linear Systems*, no. 17 in Frontiers in Applied Mathematics, SIAM, Philadelphia, 1997.

[129] A. Griewank, *On automatic differentiation*, in Mathematical Programming: Recent Developments and Applications, M. Iri and K. Tanabe, eds., Dordrecht, 1989, Kluwer, pp. 83–108.

[130] A. Griewank and G. F. Corliss, eds., *Automatic Differentiation of Algorithms: Theory, Implementation, and Application*, Philadelphia, 1991, SIAM.

[131] A. Griewank and P. L. Toint, *Local convergence analysis for partitioned quasi-Newton updates*, Numer. Math., 39 (1982), pp. 429–448.

[132] ——, *On the unconstrained optimization of partially separable functions*, in Nonlinear Optimization, M. J. D. Powell, ed., London, 1982, Academic Press.

[133] ——, *Partitioned variable metric methods for large sparse optimization problems*, Numer. Math, 39 (1982), pp. 119–137.

[134] L. Grippo and S. Lucidi, *A globally convergent version of the polak-ribière conjugate gradient method*, Math. Programming, 78 (1997), pp. 375–392.

[135] W. A. Gruver and E. Sachs, *Algorithmic Methods In Optimal Control*, Pitman, London, 1980.

[136] W. Hackbusch, *On the fast solving of parabolic boundary control problems*, SIAM J. Control Optim., 17 (1979), pp. 231–244.

[137] ——, *Optimal $H^{p,p/2}$ error estimates for a parabolic Galerkin method*, SIAM J. Numer. Anal., 18 (1981), pp. 681–692.

[138] ——, *Multi-Grid Methods and Applications*, vol. 4 of Springer Series in Computational Mathematics, Springer-Verlag, New York, 1985.

[139] W. W. HAGER, *Rates of convergence for discrete approxiamtions to unconstrained optimal control problems*, SIAM J. Numer. Anal., 13 (1976), pp. 449–472.

[140] M. HEINKENSCHLOSS, M. ULBRICH, AND S. ULBRICH, *Superlinear and quadratic convergence of affine scaling interior-point newton methods for problems with simple bounds and without strict complementarity assumption*, Mathematical Programming, 86 (1999), pp. 615–635.

[141] M. R. HESTENES AND E. STEIFEL, *Methods of conjugate gradient for solving linear systems*, J. of Res. Nat. Bureau Standards, 49 (1952), pp. 409–436.

[142] M. R. HOARE AND P. PAL, *Physical cluster mechanics: statics and energy surfaces for monoatomic systems*, Adv. Phys, 20 (1971), pp. 161–196.

[143] J. H. HOLLAND, *Genetic algorithms and the optimal allocation of trials*, SIAM J. Comput., 2 (1973).

[144] ——, *Adaption in Natural and Artificial Systems*, Univ. of Mich. Press, Ann Arbor, Mich., 1975.

[145] R. HOOKE AND T. A. JEEVES, *'Direct search' solution of numerical and statistical problems*, Journal of the Association for Computing Machinery, 8 (1961), pp. 212–229.

[146] R. HORST, P. M. PARDOLOS, AND N. V. THOAI, *Introduction to Global Optimization*, Kluwer Academic Publishers, Dordrecht, The Netherlands, 1995.

[147] W. HUYER AND A. NEUMAIER, *Global optimization by multilevel coordinate search*. Institut für Mathematik, Universität Wien, preprint, 1997.

[148] D. M. HWANG AND C. T. KELLEY, *Convergence of Broyden's method in Banach spaces*, SIAM J. Optim., 2 (1992), pp. 505–532.

[149] G. JASEN, *Investment dartboard: Pros and dart throwers both lose money*, 1997. Your Money Matters, Wall Street Journal, May 7.

[150] D. R. JONES, C. C. PERTTUNEN, AND B. E. STUCKMAN, *Lipschitzian optimization without the Lipschitz constant*, J. Optim. Theory Appl., 79 (1993), pp. 157–181.

[151] L. KANTOROVICH AND G. AKILOV, *Functional Analysis*, Pergamon Press, New York, second ed., 1982.

[152] R. B. KEARFOTT, *Rigorous Global Search: Continuous Problems*, Kluwer, Dordrecht, 1966.

[153] C. T. KELLEY, *Identification of the support of nonsmoothness*, in Large Scale Optimization: State of the Art, W. W. Hager, D. W. Hearn, and P. Pardalos, eds., Boston, 1994, Kluwer Academic Publishers B.V., pp. 192–205.

[154] ——, *Iterative Methods for Linear and Nonlinear Equations*, no. 16 in Frontiers in Applied Mathematics, SIAM, Philadelphia, 1995.

[155] ——, *Detection and remediation of stagnation in the Nelder-Mead algorithm using a sufficient decrease condition*, SIAM J. Optim., 10 (1999), pp. 43–55.

[156] C. T. Kelley, C. T. Miller, and M. D. Tocci, *Termination of Newton/chord iterations and the method of lines*, SIAM J. Sci. Comp., 19 (1998), pp. 280–290.

[157] C. T. Kelley and E. W. Sachs, *Applications of quasi-Newton methods to pseudoparabolic control problems*, in Optimal Control of Partial Differential Equations II - Theory and Applications, May, 1986, Basel, 1987, Birkhäuser.

[158] ——, *Quasi-Newton methods and unconstrained optimal control problems*, SIAM J. Control and Optimization, 25 (1987), pp. 1503–1517.

[159] ——, *A pointwise quasi-Newton method for unconstrained optimal control problems*, Numer. Math., 55 (1989), pp. 159–176.

[160] ——, *Pointwise Broyden methods*, SIAM J. Optim., 3 (1993), pp. 423–441.

[161] ——, *Multilevel algorithms for constrained compact fixed point problems*, SIAM J. Sci. Comp., 15 (1994), pp. 645–667.

[162] ——, *Local convergence of the symmetric rank-one iteration*, Computational Optimization and Applications, 9 (1998), pp. 43–63.

[163] ——, *A trust region method for parabolic boundary control problems*, SIAM J. Optim., 9 (1999), pp. 1064–1081.

[164] C. T. Kelley, E. W. Sachs, and B. Watson, *A pointwise quasi-Newton method for unconstrained optimal control problems, II*, J. Optim. Theory Appl., 71 (1991), pp. 535–547.

[165] H. Khalfan, R. H. Byrd, and R. B. Schnabel, *A theoretical and experimental study of the symmetric rank one update*, SIAM J. Optim., 3 (1993), pp. 1–24.

[166] S. Kirkpatrick, C. D. Geddat, and M. P. Vecchi, *Optimization by simulated annealing*, Science, 220 (1983), pp. 671–680.

[167] J. R. Koehler and A. B. Owen, *Computer experiments*, in Handbook of Statistics, Volume 13, S. Shosh and C. R. Rao, eds., New York, 1996, Elsevier, pp. 261–308.

[168] J. Kostrowicki and L. Piela, *Diffusion equation method of global minimization: Performance for standard test functions*, J. Optim. Theory Appl., (1991), pp. 269–284.

[169] J. C. Lagarias, J. A. Reeds, M. H. Wright, and P. E. Wright, *Convergence properties of the Nelder-Mead simplex algorithm in low dimensions*, SIAM J. Optim., 9 (1998), pp. 112–147.

[170] E. B. Lee and L. Markus, *Foundations of Optimal Control Theory*, J. Wiley, New York, London, Sydney, 1967.

[171] C. Lemaréchal, *A view of line searches*, in Optimization and Optimal Control, Auslander, Oettli, and Stoer, eds., no. 30 in Lecture Notes in Control and Information Sciences, Berlin, 1981, Springer Verlag, pp. 59–78.

[172] K. Levenberg, *A method for the solution of certain nonlinear problems in least squares*, Quart. Appl. Math., 4 (1944), pp. 164–168.

[173] R. M. Lewis and V. Torczon, *Rank ordering and positive bases in pattern search algorithms*, Tech. Rep. 96-71, Institute for Computer Applications in Science and Engineering, December 1996.

[174] ——, *Pattern search algorithms for bound constrained minimization*, SIAM J. Optim., 9 (1999), pp. 1082–1099.

[175] ——, *Pattern search algorithms for linearly constrained minimization*, SIAM J. Optim., 10 (2000), pp. 917–941.

[176] D. C. Liu and J. Nocedal, *On the limited memory BFGS method for large-scale optimization*, Math. Prog., 43 (1989), pp. 503–528.

[177] R. B. Long and W. C. Thacker, *Data assimilation into a numerical equatorial ocean model, part 2: Assimilation experiments*, Dyn. Atmos. Oceans, 13 (1989), pp. 465–477.

[178] E. M. Lowndes, *Vehicle Dynamics and Optimal Design*, PhD thesis, North Carolina State University, Raleigh, North Carolina, 1998.

[179] S. Lucidi and M. Sciandrone, *On the global convergence of derivative free methods for unconstrained optimization*. Reprint, Università di Roma "La Sapienza", Dipartimento di Informatica e Sistemistica, 1997.

[180] D. G. Luenberger, *Linear and Nonlinear Programming*, Addison-Wesley, London, 1984.

[181] J. N. Lyness and C. B. Moler, *Numerical differentiation of analytic functions*, SIAM J. Numer. Anal., 4 (1967), pp. 202–210.

[182] C. D. Maranas and C. A. Floudas, *A global optimization method for Weber's problem*, in Large Scale Optimization: State of the Art, W. W. Hager, D. W. Hearn, and P. Pardalos, eds., Boston, 1994, Kluwer Academic Publishers B.V., pp. 259–293.

[183] D. W. Marquardt, *A algorithm for least squares estimation of nonlinear parameters*, SIAM J., 11 (1963), pp. 431–441.

[184] J. M. Martinez, *Quasi-Newton methods for solving underdetermined nonlinear simultaneous equations*, J. Comp. Appl. Math., 34 (1991), pp. 171–190.

[185] E. S. Marwil, *Exploiting sparsity in Newton-type methods*, PhD thesis, Cornell University, 1978.

[186] H. Matthies and G. Strang, *The solution of nonlinear finite element equations*, International Journal of Numerical Methods in Engineering, 14 (1979), pp. 1613–1626.

[187] D. Q. Mayne and E. Polak, *Nondifferential optimization via adaptive smoothing*, J. Optim. Theory Appl., 43 (1984), pp. 601–613.

[188] K. I. M. McKinnon, *Convergence of the Nelder-Mead simplex method to a non-stationary point*, SIAM J. Optim., 9 (1999), pp. 148–158.

[189] E. H. Moore, *General Analysis*, 1935. Mem. Am. Phil. Soc. I.

[190] J. J. Moré, *The Levenberg-Marquardt algorithm: implementation and theory*, in Numerical Analysis, G. A. Watson, ed., no. 630 in Lecture Notes in Mathematics, Berlin, 1977, Springer-Verlag, pp. 105–116.

[191] ——, *Trust regions and projected gradients*, in System Modelling and Optimization, vol. 113 of Lecture Notes in Control and Information Sciences, Berlin, 1988, Springer Verlag, pp. 1–13.

[192] J. J. MORÉ AND D. C. SORENSEN, *Computing a trust region step*, SIAM J. Sci. Statist. Comput., 4 (1883), pp. 553–572.

[193] J. J. MORÉ AND D. J. THUENTE, *Line search algorithms with guaranteed sufficient decrease*, ACM Transactions on Mathematical Software, 20 (1994), pp. 286–307.

[194] J. J. MORÉ AND G. TORALDO, *On the solution of large quadratic programming problems with bound constraints*, SIAM J. Optim., 1 (1991), pp. 93–113.

[195] J. J. MORÉ AND S. J. WRIGHT, *Optimization Software Guide*, no. 14 in SIAM Frontiers in Applied Mathematics, SIAM, Philadelphia, 1993.

[196] J. J. MORÉ AND Z. WU, *Global continuation for distance geometry problems*, SIAM J. Optim., 7 (1997), pp. 814–836.

[197] W. MURRAY AND M. L. OVERTON, *Steplength algorithms for minimizing a class of nondifferentiable functions*, Computing, 23 (1979), pp. 309–331.

[198] S. G. NASH, *Newton-type minimization via the Lanczos method*, SIAM J. Numer. Anal., 21 (1984), pp. 770–789.

[199] ——, *Preconditioning of truncated Newton methods*, SIAM J. Sci. Statist. Comput., 6 (1985), pp. 599–616.

[200] J. L. NAZARETH, *A relationship between the BFGS and conjugate gradient algorithm and its implications for new algorithms*, SIAM J. Numer. Anal., 16 (1979), pp. 794–800.

[201] ——, *Conjugate gradient methods less dependent on conjugacy*, SIAM Review, 28 (1986), pp. 501–512.

[202] ——, *A view of conjugate gradient-related algorithms for nonlinear optimization*, in Adams and Nazareth [2], pp. 149–164.

[203] J. A. NELDER, 1998. Private Communication.

[204] J. A. NELDER AND R. MEAD, *A simplex method for function minimization*, Comput. J., 7 (1965), pp. 308–313.

[205] A. NEUMAIER, *On convergence and restart conditions for a nonlinear conjugate gradient method*. Institut für Mathematik, Universität Wien, preprint, 1997.

[206] J. NOCEDAL, *Updating quasi-Newton matrices with limited storage*, Math. Comp, 35 (1980), pp. 773–782.

[207] ——, *Theory of algorithms for unconstrained optimization*, Acta Numerica, 1 (1991), pp. 199–242.

[208] ——, *Conjugate gradient methods and nonlinear optimization*, in Adams and Nazareth [2], pp. 9–23.

[209] J. NOCEDAL AND Y. YUAN, *Combining trust region and line search techniques*, Tech. Rep. OTC 98/04, Optimization Technology Center, Northwestern University, 1998.

[210] J. A. NORTHBY, *Structure and binding of Lennard-Jones clusters:* $13 \leq n \leq 147$, J. Chem. Phys., 87 (1987), pp. 6166–6177.

[211] J. M. ORTEGA AND W. C. RHEINBOLDT, *Iterative Solution of Nonlinear Equations in Several Variables*, Academic Press, New York, 1970.

[212] R. PENROSE, *A generalized inverse for matrices*, Proc. Cambridge Phil. Soc., 51 (1955), pp. 406–413.

[213] L. R. PETZOLD, *A description of DASSL: a differential/algebraic system solver*, in Scientific Computing, R. S. Stepleman et al., ed., North Holland, Amsterdam, 1983, pp. 65–68.

[214] S. A. PIYAWSKII, *An algorithm for finding the absolute extremum of a function*, USSR Comp. Math. and Math. Phys., 12 (1972), pp. 57–67.

[215] E. POLAK AND G. RIBIÈRE, *Note sur la convergence de methodes de directions conjugées*, Rev Française Informat Recherche Operationelle, 3e Année, 16 (1969), pp. 35–43.

[216] B. T. POLYAK, *The conjugate gradient method in extremal problems*, USSR Comp. Math. and Math. Phys., 9 (1969), pp. 94–112.

[217] M. J. D. POWELL, *A FORTRAN subroutine for unconstrained minimization, requiring first derivatives of the objective function*, Tech. Rep. AERE-R, 6469, Mathematics Brance, A. E. R. E. Harwell, Berkshire, England, 1970.

[218] ——, *A hybrid method for nonlinear equations*, in Numerical Methods for Nonlinear Algebraic Equations, P. Rabinowitz, ed., Gordon and Breach, New York, 1970, pp. 87–114.

[219] ——, *A new algorithm for unconstrained optimization*, in Nonlinear Programming, J. B. Rosen, O. L. Mangasarian, and K. Ritter, eds., Academic Press, New York, 1970, pp. 31–65.

[220] ——, *Convergence properties of a class of minimization algorithms*, in Nonlinear Programming 2, O. L. Mangasarian, R. R. Meyer, and S. M. Robinson, eds., Academic Press, New York, 1975, pp. 1–27.

[221] ——, *Some global convergence properties of a variable metric algorithm without exact line searches*, in Nonlinear Programming, R. Cottle and C. Lemke, eds., American Mathematical Society, Providence, RI, 1976, pp. 53–72.

[222] ——, *Nonconvex minimization calculations and the conjugate gradient method*, Lecture Notes in Mathematics 1066, Springer Verlag, Berlin, (1984), pp. 122–141.

[223] ——, *On the global convergence of trust region algorithms for unconstrained minimization*, Math. Prog., 29 (1984), pp. 297–303.

[224] ——, *Convergence properties of algorithms for nonlinear optimization*, SIAM Review, 28 (1986), pp. 487–500.

[225] ——, *How bad are the BFGS and DFP methods when the objective function is quadratic*, Math. Prog., 34 (1986), pp. 34–47.

[226] ——, *Update conjugate directions by the BFGS formula*, Math. Prog, 38 (1987), pp. 29–46.

[227] ——, *Direct search algorithms for optimization calculations*, Acta Numerica, 7 (1998), pp. 287–336.

[228] K. Radhakrishnan and A. C. Hindmarsh, *Description and use of LSODE, the Livermore solver for ordinary differential equations*, Tech. Rep. URCL-ID-113855, Lawrence Livermore National Laboratory, December 1993.

[229] W. Rudin, *Principles of Mathematical Analysis*, McGraw-Hill, New York, 1953.

[230] J. Sacks, W. J. Welch, T. J. Mitchell, and H. P. Wynn, *Desings and analysis of computer experiments*, Stat. Sci., 4 (1989), pp. 409–435.

[231] R. B. Schnabel and E. Eskow, *A new modified Cholesky factorization*, SIAM J. Sci. Statist. Comput., 11 (1990), pp. 1136–1158.

[232] G. A. Schultz, R. B. Schnabel, and R. H. Byrd, *A family of trust-region-based algorithms for unconstrained minimization with strong global convergence properties*, SIAM J. Numer. Anal., 22 (1985), pp. 47–67.

[233] V. E. Shamanskii, *A modification of Newton's method*, Ukran. Mat. Zh., 19 (1967), pp. 133–138. (in Russian).

[234] L. F. Shampine, *Implementation of implicit formulas for the solution of ODEs*, SIAM J. Sci. Statist. Comput., 1 (1980), pp. 103–118.

[235] ——, *Numerical Solution of Ordinary Differential Equations*, Chapman and Hall, New York, 1994.

[236] L. F. Shampine and M. W. Reichelt, *The MATLAB ODE suite*, SIAM J. Sci. Comput., 18 (1997), pp. 1–22.

[237] D. F. Shanno, *Conditioning of quasi-Newton methods for function minimization*, Math. Comp., 24 (1970), pp. 647–657.

[238] ——, *On the variable metric methods for sparse Hessians*, Math. Comp, 34 (1980), pp. 499–514.

[239] J. Sherman and W. J. Morrison, *Adjustment of an inverse matrix corresponding to changes in the elements of a given column or a given row of the original matrix (abstract)*, Ann. Math. Stat., 20 (1949), p. 621.

[240] ——, *Adjustment of an inverse matrix corresponding to a change in one element of a given matrix*, Ann. Math. Stat., 21 (1950), pp. 124–127.

[241] B. Shubert, *A sequential method seeking the global maximum of a function*, SIAM J. Numer. Anal., 9 (1972), pp. 379–388.

[242] D. C. Sorensen, *Newton's method with a model trust region modification*, SIAM J. Numer. Anal., 19 (1982), pp. 409–426.

[243] ——, *Minimization of a large-scale quadratic function subject to a spherical constraint*, SIAM J. Optim., 7 (1997), pp. 141–162.

[244] W. Spendley, G. R. Hext, and F. R. Himsworth, *Sequential application of simplex designs in optimisation and evolutionary operation*, Technometrics, 4 (1962), pp. 441–461.

[245] W. SQUIRE AND G. TRAPP, *Using complex variables to estimate derivatives of real functions*, SIAM Review, 40 (1998), pp. 110–112.

[246] M. SRINIVAS AND L. M. PATNAIK, *Genetic algorithms: a survey*, Computer, 27 (1994), pp. 17–27.

[247] T. STEIHAUG, *The conjugate gradient method and trust regions in large scale optimization*, SIAM J. Numer. Anal., 20 (1983), pp. 626–637.

[248] C. P. STEPHENS AND W. BARITOMPA, *Global optimization requires global information*, J. Optim. Theory Appl., 96 (1998), pp. 575–588.

[249] G. W. STEWART, *Introduction to matrix computations*, Academic Press, New York, 1973.

[250] D. STONEKING, G. BILBRO, R. TREW, P. GILMORE, AND C. T. KELLEY, *Yield optimization using a GaAs process simulator coupled to a physical device model*, IEEE Transactions on Microwave Theory and Techniques, 40 (1992), pp. 1353–1363.

[251] D. E. STONEKING, G. L. BILBRO, R. J. TREW, P. GILMORE, AND C. T. KELLEY, *Yield optimization using a GaAs process simulator coupled to a physical device model*, in Proceedings IEEE/Cornell Conference on Advanced Concepts in High Speed Devices and Circuits, IEEE, 1991, pp. 374–383.

[252] K. TABABE, *Continuous Newton-Raphson method for solving an underdetermined system of nonlinear equations*, Journal of Nonlinear Analysis, Theory Methods and Applications, 3 (1979), pp. 495–503.

[253] ——, *Global analysis of continuous analogs of the Levenberg-Marquardt and Newton-Raphson methods for solving nonlinear equations*, Ann. Inst. Stat. Math., part B, 37 (1985), pp. 189–203.

[254] T. TIAN AND J. C. DUNN, *On the gradient projection method for optimal control problems with nonnegative L2 inputs*, SIAM J. Control Optim., 32 (1994), pp. 516–537.

[255] P. L. TOINT, *On sparse and symmetric matrix updating subject to a linear equation*, Math. Comp., 31 (1977), pp. 954–961.

[256] ——, *On the superlinear convergence of an algorithm for solving a sparse minimization problem*, SIAM J. Numer. Anal., 16 (1979), pp. 1036–1045.

[257] ——, *Towards an efficient sparsity exploiting Newton method for minimization*, in Sparse Matrices and Their Uses, I. S. Duff, ed., London, 1981, Academic Press, pp. 57–88.

[258] ——, *On large scale nonlinear least squares calculations*, SIAM J. Scientific and Statistical Computing, 8 (1987), pp. 416–435.

[259] ——, *Global convergence of a class of trust–region methods for nonconvex minimization in Hilbert space*, IMA J. Numer. Anal., 8 (1988), pp. 231–252.

[260] V. TORCZON, *Multidirectional Search*, PhD thesis, Rice University, Houston, Texas, 1989.

[261] ——, *On the convergence of the multidimensional direct search*, SIAM J. Optim., 1 (1991), pp. 123–145.

[262] ——, *On the convergence of pattern search algorithms*, SIAM J. Optim., 7 (1997), pp. 1–25.

[263] ——, 1998. Private communication.

[264] L. N. Trefethen and D. Bau, *Numerical Linear Algebra*, SIAM, Philadelphia, 1996.

[265] P. van Laarhoven and E. Aarts, *Simulated annealing, theory and practice*, Kluwer, Dordrecht, 1987.

[266] L. N. Vicente, *Trust-Region Interior-Point Algorithms for a Class of Nonlinear Programming Problems*, PhD thesis, Rice University, Houston, Texas, 1996.

[267] F. H. Walters, L. R. Parker, S. L. Morgan, and S. N. Demming, *Sequential Simplex Optimization*, CRC Press, Boca Raton, 1991.

[268] B. Watson, *Quasi-Newton Methoden für Minimierungsprobleme mit strukturierter Hesse-Matrix*, Diploma Thesis, Universität Trier, 1990.

[269] J. Werner, *Über die globale konvergenz von Variable-Metric Verfahren mit nichtexakter Schrittweitenbestimmung*, Numer. Math, 31 (1978), pp. 321–334.

[270] T. A. Winslow, R. J. Trew, P. Gilmore, and C. T. Kelley, *Doping profiles for optimum class B performance of GaAs mesfet amplifiers*, in Proceedings IEEE/Cornell Conference on Advanced Concepts in High Speed Devices and Circuits, IEEE, 1991, pp. 188–197.

[271] ——, *Simulated performance optimization of GaAs MESFET amplifiers*, in Proceedings IEEE/Cornell Conference on Advanced Concepts in High Speed Devices and Circuits, IEEE, 1991, pp. 393–402.

[272] P. Wolfe, *Convergence conditions for ascent methods*, SIAM Review, 11 (1969), pp. 226–235.

[273] ——, *Convergence conditions for ascent methods II: some corrections*, SIAM Review, 13 (1971), pp. 185–188.

[274] M. H. Wright, *Direct search methods: Once scorned, now respectable*, in Numerical Analysis 1995 (Proceedings of the 1995 Dundee Bienneal Conference in Numerical Analysis, D. F. Griffiths and G. A. Watson, eds., Harlow, United Kingdom, 1996, Addison Wesley Longman, pp. 191–208.

[275] S. J. Wright, *Compact storage of Broyden-class quasi-Newton matrices*. Argonne National Laboratory, preprint, 1994.

[276] S. J. Wright and J. N. Holt, *An inexact Levenbert-Marquardt method for large sparse nonlinear least squares*, J. Austral. Math. Soc. Ser. B, 26 (1985), pp. 387–403.

[277] Z. Wu, *The effective energy transformation scheme as a special continuation approach to global optimization with application to molecular confirmation*, SIAM J. Optim., (1996), pp. 748–768.

[278] T. J. Ypma, *The effect of rounding errors on Newton-like methods*, IMA J. Numer. Anal., 3 (1983), pp. 109–118.

[279] S. K. Zavriev, *On the global optimization properties of finite-difference local descent algorithms*, J. Global Optimization, 3 (1993), pp. 67–78.

[280] C. Zhu, R. H. Byrd, P. Lu, and J. Nocedal, *L-BFGS-B - FORTRAN subroutines for large-scale bound constrained optimization*, ACM Transactions on Mathematical Software, 23 (1997), pp. 550–560.

Index